Around the World in 80 Ways

Stephen Webb

Around the World in 80 Ways

Exploring Our Planet Through Maps and Data

Stephen Webb
DCQE
University of Portsmouth
Portsmouth, UK

ISBN 978-3-031-02442-9   ISBN 978-3-031-02440-5   (eBook)
https://doi.org/10.1007/978-3-031-02440-5

This Springer imprint is published by the registered company Springer Nature Switzerland AG
The registered company address is: Gewerbestrasse 11, 6330 Cham, Switzerland

To Chris Caron, Angela Lahee, and Lisa Scalone
who have always been a delight to work with

Acknowledgments

I would like to thank the Open Source and the Creative Commons communities. Without their generosity, and the giving of their time and knowledge, the writing of this book would have been a difficult and arduous task. Thanks to them, this was a fun project for me. I would also like to thank Alina Maniar and Carl Berndtsson for kindly allowing the reproduction of photographs.

I am particularly grateful to Brian Clegg for his extensive, expert, and insightful comments on an early draft of the manuscript. Any merits this book might possess will be in large part due to Brian's input. The mistakes, of course, are all my own. Lisa Scalone has once again provided brilliant editorial advice.

Finally, as always, I would like to thank Heike and Jessica for everything.

Contents

Mapping the World

Over the years I've had discussions with people who hold, shall we say, opinions that contradict mainstream scientific thought. Often, I felt, my interlocutors—anti-vaxxers, climate-change deniers, flat-Earthers, and others who discern conspiracy where the rest of us see science—refused to accept that human activity can impact the globe ("we're just one species; how can we cause climate change?") or even that there *is* a globe ("if Earth is curved, how come we can sometimes see very distant skyscrapers?"). How can they be so *ignorant* of the world, I'd wonder.

And then I read Hans Rosling's book *Factfulness*, the introduction to which contains a dozen simple questions about the world. Alongside each question Rosling offered three possible answers. I got four answers correct: the same as a chimpanzee would score were it to point randomly at the page. I consoled myself with the thought that Rosling posed those same questions to almost 12,000 people in 14 countries and not a single person got all of them right. The average score was just two correct answers (worse, then, than a chimpanzee) and 1800 people got every question wrong. It seems that *all* of us, Nobel prize winners included, are ignorant about many aspects of our world. But if my own knowledge was shaky, how could I in good conscience argue with the anti-vaxxers, the climate-change deniers, and the flat-Earthers?

I got into the habit of testing my intuition against the most robust data I could find. So I asked myself questions such as: What is the commonest cause of death in different countries? Do different nationalities search Google for different terms or are people across the globe interested in the same things? Which nations generate the most trash? At the same time I discovered that, using the power of modern GIS software, it is easy to create maps of these data—and a map is far easier to grasp than a table of numbers.

After making a few maps I hoped they would help me prove my case when I got into discussions with people who have an allergy to science. I didn't realise that some people set the bar low when it comes to evidence that might support their beliefs but impossibly high for evidence that might challenge them. My charts changed no minds.

Nevertheless, I enjoyed the process of making the maps—so I made more of them. I made 80, in fact, some quirky, others heartbreaking. They fall naturally into eight themes, and I've collected them here. However, as anti-vaxxers, climate-change deniers, and flat-Earthers would quickly point out, my maps deceive. How do I deceive thee? Let me count the ways...

1 Map Projections

All maps project the surface of a three-dimensional sphere (in other words, Earth's surface) onto a two-dimensional space (such as the page of a book). It's mathematically impossible to do that without distorting something. This fact is presumably of no interest to flat-Earthers, but it should bother the rest of us. One can project Earth's curved surface onto a plane in an infinite number of different ways, and all of them distort the 'truth' in some way. The question is: what distortion are we willing to accept and what features do we want to preserve?

A map has four basic characteristics—area, direction, distance, and shape. Map projections differ in how they try to preserve these characteristics. In my primary school the map adorning the wall was a Mercator projection. A Mercator map is useful for sailors: it preserves directions, so any course of constant bearing appears as a straight-line segment on the map. The downside is that, while areas are accurate close to the equator, areas inflate as one heads towards the poles. For years, thanks to the Mercator map hanging on my school wall, I believed Antarctica is Earth's largest continent and Greenland rivals America in size.

© The Author(s), under exclusive license to Springer Nature Switzerland AG 2023
S. Webb, *Around the World in 80 Ways*, https://doi.org/10.1007/978-3-031-02440-5_1

The American inventor Buckminster Fuller wanted a map of the world that better preserved the relative size of areas and the shapes of areas. He projected the world map onto the surface of an icosohedron, a three-dimensional object consisting of 20 equilateral triangles, and then cut the icosohedron so it could be laid out flat. His Dymaxion map (see Fig. 1, left), better known as the Fuller projection, butchers the world; good luck using it for navigation. But since one can cut the projection in many different ways, one can use it to illustrate themes that are difficult to show with other map projections.

Or consider the Werner projection (see Fig. 1, right), developed by the German mathematician Johannes Werner in the sixteenth century, which turns the world into a heart. I can think of few practical uses for a map based on the Werner projection. But the projection is not *wrong*. And isn't there something uplifting about seeing the world in the shape of a heart?

There is no 'right' map projection. From the infinite options on offer, you simply use the one that suits your task.

For the purposes of this book a map projection that seeks to retain the relative size of areas makes sense. I have chosen the Equal Earth projection, developed in 2018 by three geodetic engineers, Bojan Šavrič, Tom Patterson, and Bernhard Jenny. On an Equal Earth map Greenland no longer competes against America in size. Be aware, however, that this projection distorts and stretches shapes, directions, and distances north–south. I believe these maps can be *useful* and I try throughout to be *honest*, but the maps are not *truthful*—nor can they be. Lie number one.

2   Map Conventions

We bring our prejudices to the act of making and reading maps. For example, the prime meridian, the line of longitude 0°, runs through Greenwich. My maps are thus centred on a line that takes in London. But that particular definition of the prime meridian is an accident of history. In the latter half of the nineteenth century, much of the world's commerce relied on sea-charts that had a prime meridian centred on Greenwich; furthermore the USA used it as the basis of its national system of time zones. So formalising Greenwich as the centre of world time, something agreed at the International Meridian Conference in 1884, was convenient for the largest number of people *at that time*. Our maps do not have to look that way. The map that used to hang on my old school wall could just as reasonably have been centred on a line taking in New York, or Moscow, or Shanghai. In Fig. 2, the world map is centred on the 150th meridian east, a line that takes in Queensland and New South Wales in Australia.

Although this map might seem unfamiliar, it is recognisable. But another convention is that north should appear at the top of a map. We could just as easily adopt the opposite convention. This 'upside-down' view of the world (see Fig. 3) is *entirely* unfamiliar. It looks *wrong*. And yet it is as valid as our usual representation. If we wished we could represent our maps with east, say, at the top. In medieval Europe, before explorers adopted the magnetic compass, most mapmakers did just that: the rising of the sun provided an important bearing.

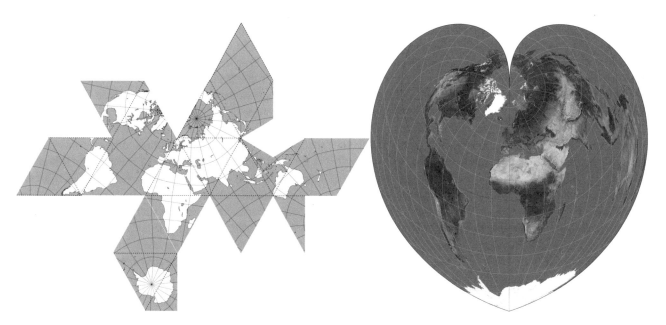

Fig. 1 Left: the Dymaxion, or Fuller, projection of the world centred on Europe and Africa. (Thomasee73, CC BY-SA 4.0). Right: the Werner projection of the world, with imagery derived from NASA's Blue Marble summer month composite. (Strebe, CC BY-SA 3.0)

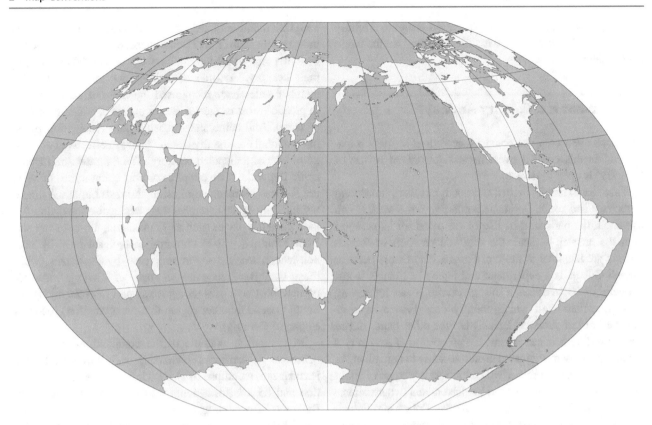

Fig. 2 The world map centred on the 150 meridian east, which takes in the Pacific Ocean. This map uses the Winkel III projection. (Milenioscuro, CC BY-SA 3.0)

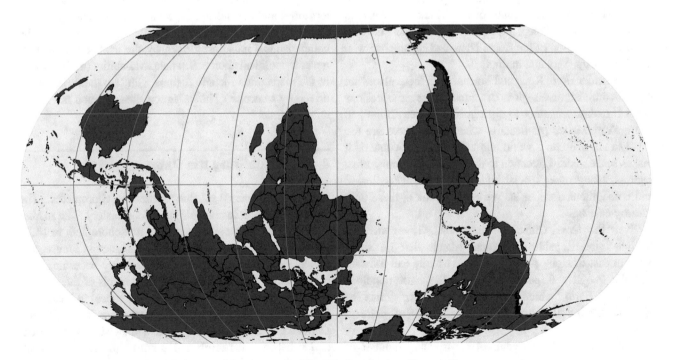

Fig. 3 The world map with south at the top. This map uses the Equal Earth projection. (Own work)

If we never stop to note that our maps enshrine arbitrary conventions then we will find it difficult to ever see the world anew. Lie number two.

3   What Is a Country Anyway?

My maps often compare countries according to some statistic. But in this context even the straightforward notion of 'country' is open to debate.

Take, for example, French Guiana. It has Brazil bordering to the east and south, Suriname bordering to the west. If you ask about the pattern of its flag or the colour of its passport then the answer is clear: the flag and passport of French Guiana are identical to those of France. And that is because French Guiana is a department and region of France; it just happens to be overseas. (This gives rise to an interesting trivia question: with which country does France have its longest border? Answer: Brazil.) On the other hand, if you happen to be interested in, say, the proportion of a country's land covered by forest then it makes no sense to argue that the figure for mainland France (which has forestry cover of about 30%) should be carried over to French Guiana (where forests cover about 99% of the land). So whether France and French Guiana should be considered identical depends upon context. It is another choice.

Or consider the UK. In many cases of interest we gain insight by comparing the figures for England, Scotland, Wales, and Northern Ireland. But differences between Northern Ireland and Wales, say, are hard to perceive on a world map. Besides, combined figures for the UK are often easier to obtain. So UK it is. Except there are occasions where the data *require* a separation of the Home Nations: England has won the World Cup, Scotland has not.

Or consider Hong Kong and Macau. Both these places are special administrative regions of China. There are occasions when it makes sense not to distinguish between China and Hong Kong/Macau (particularly when these places are too small to show up on a world map). But in some contexts it makes sense to draw attention to these special administrative regions. Many other special regions, dependent territories, and autonomous areas exist. We need to look at these on a case-by-case basis.

Even the term 'China' is ambiguous. There are two 'Chinas': the Republic of China, islands that lie about 800 km east of Hong Kong, which we usually call Taiwan; and the People's Republic of China, the most populous nation on Earth, which we usually refer to simply as 'China'. Behind that ambiguity sits decades of political controversy.

The control and ownership of a surprising number of places remain a matter of dispute. Kosovo, for example, is currently recognised as an independent state by 97 of the 193 UN member states (that's 50.3%)—so is Kosovo an independent state or not? Some states, recognised as independent by much of the international community, do not control their territory. At time of writing, for example, 138 UN member states have recognised the State of Palestine—but in practice most of the territory it claims is under the control of Israel. And some states receive little recognition by the international community despite being in effective control of all or part of a disputed territory. The Sahrawi Arab Democratic Republic, for example, controls about a fifth of Western Sahara (Morocco administers the rest) but at the time of writing only 40 UN member states recognise the claim. And then there are places such as Transnistria, a breakaway state from Moldova that is currently recognised only by other non-recognised states—in this case by Abkhazia (which most countries recognise as part of Georgia), Artsakh (which most countries recognise as part of Azerbaijan), and South Ossetia (another region that most countries recognise as part of Georgia).

My favourite example of the complexities involved in questions of territorial ownership is a piece of land in Kazakhstan, an ellipse of size 90 km in the east–west direction and 85 km north–south, at the centre of which is the Baikonur Cosmodrome. Although it lies well inside Kazakhstan it is, until a lease runs out in 2050, formally a part of Russia. I have chosen to leave this plot of land on my maps, not as a mountweazel but simply as a reminder that national boundaries are the product of often labyrinthine historical flows.

The maps in this book typically refer to about 250 nations, territories and dependencies—a number significantly greater than the number of member states of the United Nations. The comparisons I make are driven by the availability of data, never by political choice. But data availability (or the lack of it; the maps often include regions with 'No data') is always driven by *someone's* choice. Ignore that and you fall for lie number three.

4   Visualising the Data

Most of the maps in this book are choropleths. The choropleth, or 'colour map', has a long history. The mathematician Charles Dupin created the first choropleth in 1826. Dupin was interested in levels of literacy across France, and he shaded different regions of his country according to a colour scale running from white (high levels of literacy, the light colour symbolising 'enlightened' France) to black (low levels of literacy, the inky colour representing 'dark' France). Dupin's map had a practical use: at a glance it showed a clear divide between the north and south of the country, which hinted at a disparity of education between the two parts of

France. Once a problem becomes visible like this, politicians can start to develop policies to address the issue.

So choropleths can be valuable: they are an easy way of representing a large amount of data in a succinct, easy-to-digest, visually appealing way. And because a wide variety of data is collected at national level, it is a trivial task for a modern GIS system to spit out a choropleth illustrating that data. But one must be careful when interpreting a choropleth.

For example, suppose two countries possess the same value of the topic under discussion. Following on from Dupin, we might find they have the same rate of literacy, say. In this case we assign the same colour to both countries. But if one country covers a large area and the other is small, the larger country will appear more prominent: the eye can't help but attach greater significance to a larger block of colour.

And then there's the question of which colour to use. Quite apart from the cultural significance of colour, there is a practical point that has its roots in the technology of printing. Suppose you represent some quantity with a colour progression running from white (for small values of the quantity) to red (for large values). Well, printers will struggle to distinguish more than five or six shades on the choropleth. Furthermore, the human eye will struggle to discriminate one country from another if too many shades of red exist. (Personally, I struggle to distinguish between more than four shades of red.)

Then there are the choices one makes when classifying the data. Suppose we have some measurement, which runs from 1 to 100, on 250 countries. And suppose we want to put each country into one of five data groups, or 'bins', depending on its measurement. How should we proceed? We could ensure that we have 50 countries in each of the five bins. That seems reasonable. But, depending on the underlying distribution of the measurement, we might need bins of different sizes. We might end up with bins of size 1–2.7, 2.8–3.6, 3.7–22.1, 22.2–22.5, 22.5–100, say. Besides leading to bin sizes that are awkward to work with, it appears unfair to group a country with a measurement of 22.5 with one that measures 100. Or we could decide to have bins of equal size: 1–20; 21–40; 41–60; 61–80; 81–100. That, too, seems reasonable. But, depending on the underlying distribution of the measurement, we might end up with very different numbers of countries in the different bins: 145, 85, 17, 2, 1, say. That, too, appears unfair. Or we could try to minimise variation within each bin while also rounding the boundaries up to whole numbers for ease of reading. Those three approaches (and there are other options I haven't mentioned) would lead to maps with a different appearance in each case *but the underlying data would be the same*. Map appearance depends on our choice of how to classify the data. Remember, when you look at these maps, that one could create other valid visualisations. Otherwise you fall for lie number four.

5 Collecting and Interpreting the Data

The Covid-19 pandemic has given us all a lesson in the importance of data, but also in how difficult it can be to collect data in a consistent way and in how fiendishly hard it can be to interpret data.

The SARS-CoV-2 virus exists in order to find a host in which it can replicate and, once that host is no longer useful, move on to another. The illness caused by the virus is a by-product of that drive to replicate. So we have a situation in which the virus spreads through the population, and governments in turn respond by restricting the movement of people and by implementing public health measures (with varying degrees of rigour). Now consider how difficult it is to answer the most basic question about this virus: how deadly is the disease, Covid-19, that it causes?

Journalists often write about the case fatality rate. It *should* be easy to determine, right? You just need two numbers: the number of deaths from disease divided by the number of diagnosed cases over some period of time. But the world is not that simple.

Take the number of cases. Some people infected with the SARS-CoV-2 virus can be asymptomatic: they don't know they have had the infection. If a person is feeling fine then it's unlikely he or she would be tested for Covid-19; but that creates an uncertainty in our estimate for the number of cases. When people *are* tested some (admittedly small) proportion of tests return a wrong result because no diagnostic test is ever perfect: we get false positives and false negatives. Test results can even go missing, which again generates an uncertainty. Other factors can be at play, too: in some states of the USA, for example, politicians have asked scientists to 'massage' data in order to align with a particular view. And in many countries, the public health infrastructure is too fragile to collect reliable data. All this and more means case numbers, even confirmed case numbers, come with some uncertainty attached.

But we can at least count the number of deaths caused by Covid-19, right? Well, no. Public Health England provides a daily count of those who died within 28 days of testing positive for coronavirus, regardless of cause of death. On the other hand, particularly early on in the pandemic, the data did not include people who almost certainly died as a result of the virus but who were never tested. Even if everyone were tested with a perfectly sensitive test there would still be ambiguity: if someone dies of a stroke five weeks after testing positive, should that be counted as a Covid-19 death? Perhaps; perhaps not. It would be a judgement call for a doctor. (And I am ignoring here a death caused by a heart attack, say, that would have been prevented had the victim sought assistance—but did not do so because of worries of contracting Covid-19 in the health care system. In that case the SARS-CoV-2 virus would not be in the patient's body, but the patient died because the virus was in the body of the community.) Maybe the only way to measure the number of deaths is to count excess deaths over the most recent five-year average and attribute the excess to Covid-19. Even that approach is problematic because the response to the pandemic might have reduced deaths from other causes (fewer cars on the road, for example, means fewer road traffic fatalities).

Determining the impact of Covid-19 in a single jurisdiction is hard. The difficulty increases when one attempts to compare nations. Different countries have reported Covid-19 deaths in different ways: some countries adopted a broad definition of what constitutes a coronavirus fatality, others tried to cover up the number of deaths. Healthcare systems and social support differ between countries. Furthermore, one thing we know for certain about Covid-19 is that it kills older people more readily than it kills younger people, so comparisons should really take into account the age profile of a country.

This is not a primer on epidemiology, so I'll leave the discussion there. The point I want to make is that collecting data, even for something as straightforward as the number of people who die from a particular disease, is difficult. It can be even more difficult to divine the meaning of that data. Experts can—and often do—disagree over the meaning of data. (And bear in mind, as you read, that I often express an opinion on data in areas in which I have no particular expertise.)

This is not to say we should avoid attempts at comparing countries. In the case of Covid-19, for example, surely we *should* try to compare countries, even if we know such comparisons are imperfect, because we might be able to learn more about the virus. Nevertheless, we all need to question the collection and interpretation of data—or we fall prey to lie number five.

6   A Moment in Time

The maps in this book refer to the latest data to which I have access. But things change. Remember, as you read, that these maps represent snapshots of activity rather than immutable truths. Otherwise, lie number six will deceive you.

7   Questions, Questions

Lie number seven is subtle: even the questions we choose to ask can be a source of bias. In *Factfulness*, for example, one of Rosling's questions was: "In 1996, tigers, giant pandas, and black rhinos were all listed as endangered. How many of these three species are more critically endangered today?"

We feel better when we learn the surprising answer: "None of them". Had Rosling asked instead about South China tigers, polar bears, and northern white rhinos—well, we wouldn't feel so good. It matters, which questions we choose to ask.

* * *

Seven deadly deceptions. You might conclude that anti-vaxxers, climate-change deniers, and flat-Earthers have a point. If maps are so misleading, why bother engaging with them? Indeed, there is a deeper question here: why should any of us attempt to understand the world in this way if the truth is so elusive?

I can think of a number of reasons why the attempt is worth it.

The first is it reminds us science is not a body of facts nor a collection of truths. Rather, it is a *process*. Scientists understand that science does not achieve complete certainty, but it *is* our best route to finding robust and reliable knowledge. We need to learn to live with uncertainty. If nothing else, accepting uncertainty is healthier than having complete faith in answers that are wrong. Particularly in these times, when many of our political leaders start with a gut feeling and then look for 'alternative facts' to justify their feeling, this is an important lesson.

Even where we have uncertain data we can still learn things, still draw conclusions, still compare one part of the world with another. We don't know, for example, the precise number of judicial executions taking place in China each year. But we *do* know the number is greater than that of any other country, and we *do* know many countries have abolished the death penalty. A map of capital punishment, illustrating the number of executions by country in a given year, cannot be completely accurate. But it can still provide insights.

There are other reasons for working with these maps. It's *interesting* to ponder the location of world heritage sites, say, or the world's tallest buildings. It's *fun* to reflect on national success at the World Cup Finals, say, or the distribution of medals at the Olympics. It's *important* to contemplate different peoples' standard of living, say, or their access to electricity.

Finally, if you disagree with the maps here—perhaps you think my colour scheme is misleading, or I put the data into too many or too few bins, or the size of the bins is confusing . . . well, you can create your own! I provide details of the sources I used so you can generate different versions of any of these maps or, as new data become available, create updated versions. The creation of world maps was once a task for professional cartographers. Nowadays, as I explain in an appendix, the widespread availability of open source geographical information systems mean anyone who can use a computer can generate a choropleth. Give it a try!

The World Itself

Astronomers now know of thousands of exoplanets, planets that orbit distant stars. None of those planets are much like ours. This might be because our techniques favour the detection of giant planets and planets that orbit close to their parent star. But it is also possible that rocky, tectonically active, water-rich planets in possession of a large satellite—in other words, life-bearing places like Earth—are rare. In this first chapter we take a look at our planetary home.

1 Earth's Poles (Map 1)

Just over 4.54 billion years ago, part of a giant molecular cloud began to collapse under the force of gravity. The collapse led to the formation of a protostar, the progenitor of our Sun, around which a disc of gas and dust began to orbit. Over time, dust grains would occasionally collide and clump together, eventually forming rocks as large as 200 m. In turn, these rocks collided and formed planetesimals as large as 10 km. The collisions continued for several million years, and led to the formation of four terrestrial planets (Mercury, Venus, Earth, Mars) in the inner solar system and four giant planets (Jupiter, Saturn, Uranus, Neptune) in the outer solar system, along with numerous smaller bodies. Collisions in the solar system never stopped (they continue still) and some astronomers hypothesise that in one such collision Theia, an object the size of Mars, struck the infant Earth. That impact formed the Moon. Ever since, Earth and Moon have danced together in their yearly lap of the Sun.

If the Theia hypothesis is correct, that Moon-forming collision caused Earth's rotational axis to tilt. We feel the results to this day. Our planet's axis of rotation, which meets the surface at the North and South Poles, is tilted at just over 23° with respect to the plane of Earth's orbit around the Sun. And because this tilt is fixed, regardless of where Earth is in its annual orbit, we experience seasons. Summer in the northern hemisphere sees the south polar regions freeze in darkness; when it is summer in the southern hemisphere the north

polar regions experience the dark. In total, Earth's polar regions receive less solar radiation than its temperate and torrid zones—and so surface temperatures are lower at the poles. Over time, therefore, the polar regions develop ice caps.

As a child, I believed the ice cap at Antarctica must be Earth's biggest continent: the bottom of any map showed a sprawling land mass across Earth's full width. But Antarctica is only the fifth largest of the seven continents. Its apparent size is the result of the typical methods used to represent Earth's curvature on a flat page. As we saw in the Introduction, when you depict a curved surface on a flat map you must distort some element of the map. For most map makers, and for most map readers, it makes sense to distort the polar regions (since few people ever visit them) and maintain an accurate representation of the temperate and torrid zones (since people live there). But such distortion is a problem when a map illustrates some form of human activity because the question of what happens at the poles is irrelevant: it is not worth considering the incidence of road traffic accidents at the North Pole, say, or deaths from malaria at the South Pole. The Equal Earth projection used in this book compresses features in the north–south direction near the poles, so the polar regions do not dominate and the relative size of Antarctica is shown correctly. Since this small continent often appears in this book in grey (in order to denote 'No data') the reader's eye might slip over the area at the bottom of a map. There is then a danger that we *underestimate* the importance of the poles. The polar regions are essential to humanity's future. So, before we start focus on the rest of Earth, let's look at the polar regions. First, the North Pole.

Maps that 'look down' from above often show the North Pole as a point in the blue of the Arctic Ocean. The North Pole does indeed lay in the middle of the Arctic Ocean. The nearest land, Kaffeklubben (Coffee Club) Island, is about 690 km away just off the northern tip of Greenland. The nearest permanently inhabited settlement is 810 km away,

S. Webb, *Around the World in 80 Ways*, https://doi.org/10.1007/978-3-031-02440-5_2

1   Earth's poles

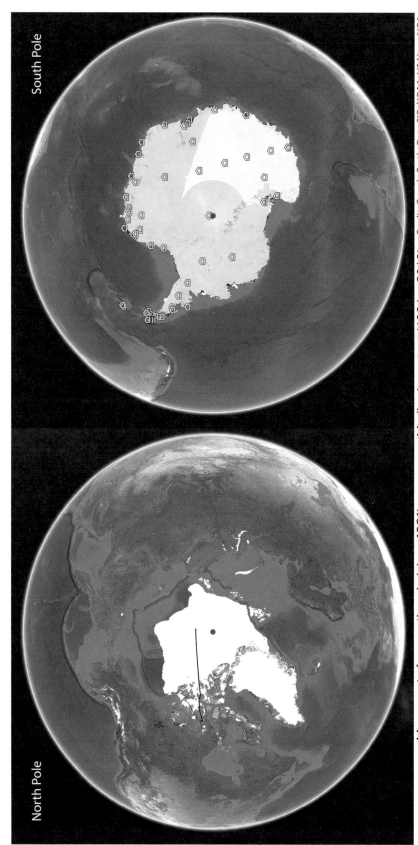

North Pole

South Pole

- Pole
- —— Movement of magnetic pole (since 1831)
- —— Movement of magnetic pole (1590 – 1831)
- —— Median ice extent (1981 – 2010)
- ◎ Scientific base

Credit: Google Earth; Data SIO, NOAA, US Navy, NGA, GEBCO, NSIDC; Image Landsat/Copernicus; Image US Geological Survey; Image PGC/NASA

Map 1

Fig. 4 To reach the North Pole with a surface ship requires something like this Russian vessel, a nuclear-powered icebreaker called the '50 Years of Victory'. The ship can break through ice up to 2.8 m thick, whether sailing backwards or forwards. Nuclear-powered submarines are another option: they can sail beneath the ice, then punch a hole through it when they surface. (Christopher Michel, CC BY 3.0)

in the Canadian territory of Nanavut. But such maps mislead because they suggest that you or I could sail across the top of the world. With a normal boat, that journey is impossible (see Fig. 4). The high Arctic waters have an almost permanent cover of shifting sea ice. Ice presents such a challenge to exploration that humans first reached the North Pole as recently as 1926, and that was in an airship. (The claims of earlier expeditions, which used wooden sleds and dog teams to attempt to reach the North Pole, have not withstood scrutiny.)

If the polar icepack were easily navigable, a European ship bound for east Asia could shave about 4000 km from its journey. Huge economic benefits would flow. Small wonder, then, that explorers began searching for a so-called 'North-west Passage' as early as the sixteenth century. As late as 1845, British explorers were still dying in the search: Sir John Franklin led an expedition of two ships, HMS *Erebus* and HMS *Terror*, neither of which returned. Today, global warming is causing the icepack to shrink and thin: the magenta line on the image shows the median ice extent for the month of October in the 30-year period 1981–2010 and the latest icepack, as can be clearly seen, is much smaller than the median. A reliable passage for commercial shipping might soon be available for a few months each year. Some economies will benefit but scientists view the creation of a Northwest Passage as being of small reward compared to the dangers associated with our climate emergency.

The North Pole is a fixed geographic feature. The North *Magnetic* Pole—the location at which a compass needle, if allowed to rotate freely, points vertically down—moves in response to happenings deep in Earth's core. In 1831 James Clark Ross became the first to reach the North Magnetic Pole. Since then the Pole has moved, and is currently drifting towards Siberia at a rate of 50 km/yr. (The black line on the image shows how it has drifted since 1831; the red line is an estimate of its position dating back to 1590.) Eventually, Earth's magnetic field will flip: north and south will swap, an event that last happened 780,000 years ago.

The South Pole, unlike the North, is not located on sea: Antarctica is a continental land mass. The ice here is about 2.7 km thick. That this ice sits on land is a cause for concern in a warming world. Arctic sea ice, when it melts, has no effect on global sea level (for the same reason an ice cube floating in a glass of water does not change the water level when it melts). A melting Antarctic would put *additional* water into the oceans and global sea level would rise. If *all* Antarctic ice were to melt, sea levels would rise by about 60 m. That extreme case will almost certainly not happen, but even a modest rise in sea levels would be enough to make some coastal cities uninhabitable (see Map 31).

Although Antarctica is a colder, harsher environment than anything found in the northern hemisphere, explorers reached the South Pole before the North. The great Norwegian polar explorer Roald Amundsen claimed the honour in 1911. Scott and four others got there 34 days after Amundsen, then perished on the return journey. (In 1914 the indefatigable explorer Ernest Shackleton attempted the first crossing of Antarctica via the South Pole. He did not make it, but his story of *Endurance* is astonishing.)

The ever-shifting nature of the Arctic icecap makes it difficult to construct permanent scientific bases there. Antarctica, in contrast, is home to about 75 permanent research bases and its population—scientists, technicians, and support staff—exceeds 1100 during winter and swells to more than 4200 during summer. One such research base, the Amundsen–Scott South Pole Station, is the southernmost structure on Earth: it lies directly at the Pole. Scientists at the Station experience just one 'day' and one 'night' during the year: for six months the Sun is above the horizon and for six months it is below. This period of prolonged darkness, combined with the dry atmosphere (Antarctica is one of the driest deserts on Earth), makes the Station a good place for astronomy. The Station is also home to some of the most esoteric experiments on the planet. IceCube, for example, hunts for neutrinos—mysterious particles which, although abundant in the universe, are reluctant to interact with the rest of the world. IceCube snares high-energy neutrinos from the depths of the cosmos. See Fig. 5.

2 Meteorites (Map 2)

Our knowledge of how Earth came into being and of how our solar system evolved comes in large part from a study of meteorites. Certain types of meteorite that we find here on Earth developed in the same dust that gave rise to the Sun and planets, in the same protoplanetary disk from which Earth formed. The thought inspires awe: a pristine meteoroid can lap the Sun without incident for four billion years until, one day, its orbit intersects that of our planet. The meteoroid blazes through the atmosphere (at which point we call it a meteor) and then a snapshot of our solar system's pre-history falls to Earth as a meteorite. You can imagine how fortunate I felt, then, when I was given a behind-the-scenes tour of the

Fig. 5 The IceCube Laboratory, at the Amundsen–Scott South Pole Station, underneath a night sky in winter. This building hosts computers that are used to collect and process data from the experiment's detectors. The detectors themselves are buried deep underneath Antarctic ice, and look 'up' through Earth's bulk in search of neutrinos. (John Hardin, CC BY 4.0)

2 Meteorites

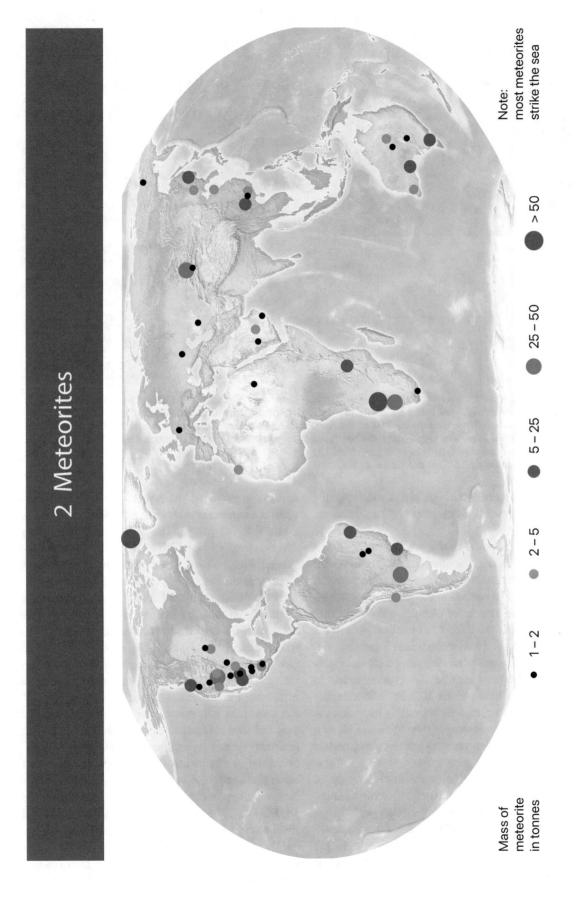

Note:
most meteorites
strike the sea

Mass of
meteorite
in tonnes

1 – 2   2 – 5   5 – 25   25 – 50   > 50

Map 2

Fig. 6 Left: the South Corston fragment of the Strathmore meteorite. The Corston fragment, which weighs just over 1 kg, landed in a field about 50 m from a farm house. The impact made a hole 15 cm deep. Right: The Easter Essendy fragment. Weighing slightly over 10 kg, this was the largest of the four meteorite fragments to be recovered. The impact made a hole 45 cm deep. The other two fragments are called Carsie and Keithick. The Carsie fragment is interesting in that an eyewitness, Mrs. Grace Walsh, saw it hit the ground. The Keithick fragment became the most famous of the four when it drilled a hole through the roof of a house occupied by the Hill family. (Geni, CC BY 4.0)

National Museums Scotland—the highlight of which was the chance to hold the largest of the four pieces of the Strathmore meteorite.

On 3 December 1917, just after 1 pm, a fireball flashed across the clear winter skies of southern Scotland. Within seconds, people in the central region of Strathmore heard an explosion as four objects crashed to the ground. Contemporary accounts describe a range of responses to the fireball. One witness said he "heard it fizzin'"; a more pompous report declared the "community [was] thrown into a state of consternation". The four objects were soon recovered, and represent parts of the largest meteorite fall ever recorded in Scotland. The piece I held weighs 10.2 kg. Mineralogists have studied the meteorite, and the prevailing hypothesis is it was once part of an ancient stony asteroid that was hit by a larger body some 468 million years ago; the impact smashed the asteroid into tiny pieces, with orbits that cross Earth's. The Strathmore meteorite is nothing special to look at, just a dull lump of rock; see Fig. 6. Nevertheless, it's quite a feeling to hold an object you know spent 468 million years in the cold of space before smashing into the cold of Scotland.

It is not every day you get to hold a space rock, but meteorites are not uncommon. As Earth follows its orbit it encounters lots of material. Most of this stuff is small: grains of dust and pea-sized bits of rock. This type of material burns up in the atmosphere—we see a "shooting star"—and never makes it to the ground. The larger bits of rock, though, can survive the shock of a high-speed journey through the atmosphere. We have two possible ways to estimate how many of them hit Earth each year. First, we can use an all-sky camera to monitor the meteorites that fall in a particular area and then (assuming all areas receive the same number of meteorites) extrapolate the total for Earth. Second, we can go to a place where there is little vegetation or erosion (a desert does nicely) and then count the number of meteorites lying around. One can estimate how long a meteorite has been there by measuring the amount of weathering. From these data one can determine how many meteorites fall in that place each year and then extrapolate the total for Earth. Although both methods generate estimates that come with large uncertainties attached to them, it seems reasonable to conclude that Earth receives between 40,000–80,000 tonnes of meteoritic mass every year. One estimate suggests between 18,000–84,000 meteorites with mass greater than 10 g hit Earth each year. The majority of meteorites fall into the ocean rather than onto land, for the simple reason that there is more than twice as much ocean cover than land cover. Ocean-falling meteorites are typically lost from view. Some

Fig. 7 A visitor centre has been created around the Hoba meteorite in Groonfontein, Namibia, which is classed as a national monument. Hoba is the largest known single-piece meteorite. (Sergio Conti, CC BY 2.0)

meteorites, though, hit land and can be recovered. The Strathmore meteorite was one such example.

The Meteoritical Society maintains a database of meteorite finds. At the time of writing the database contains 61,865 records. The location of a large fraction of these meteorites is Antarctica. This is not because Antarctica is particularly large (as we saw in Map 1, it isn't) nor is it because it attracts meteorites (it doesn't). Rather, scientists have determined that Antarctica is the ideal place to go meteorite hunting—so some of the scientific bases shown in Map 1 are home to geologists who go looking for space rocks. The conditions are ideal: the dry, cold climate preserves the meteorites that land there; the ice sheets create a "conveyor belt" that concentrates them in areas where they can be easily collected; and high-speed winds scour the ice surface and expose buried rocks.

In addition to the location field in the Meteoritical Society database, each record contains a wealth of other information. Each record, for example, contains an estimate of the mass of the meteorite. The overwhelming majority of meteorites in the database are low-mass objects, of a few grammes or less. This is exactly what one would expect: the solar system contains many more low-mass objects than high-mass objects. But the database also contains details of substantial

meteorites, objects much bigger than the Strathmore meteorite. The database has records of 53 objects with a mass greater than 1 tonne.

The most massive object in the database, weighing in at 60 tonnes, is the Hoba meteorite. Jacobus Hermanus Brits, the owner of a farm in the Otjozondjupa region of Namibia, discovered it in 1920, quite by chance, when he heard a loud metallic screech as he ploughed one of his fields. Brits had the obstruction excavated and uncovered a tablet of metal 2.7 m by 2.7 m by 0.9 m in size. Workers estimated its mass to be 66 tonnes; the combined effects of weathering, sampling, and vandalism have reduced its weight to its present 60 tonnes. Scientists soon identified the tablet as an iron meteorite that fell within the past 80,000 years. Given its weight and size this object—the most massive naturally occurring piece of iron anywhere on Earth's surface—has never moved in all its time here.

The Hoba meteorite is now a tourist attraction (see Fig. 7), receiving thousands of visitors each year, and the commonest question those tourists ask is: why is there no crater? Surely a 66 tonne lump of metal barrelling into Earth's surface would leave a sizeable hole? The answer is simple: our planet's atmosphere, acting on the object's unusually flat shape, slowed the meteorite's descent. When the iron slab crashed into Hoba it was moving merely at terminal velocity. If any eyewitnesses were around they would have heard a bang, but the meteorite fell too slowly to cause a significant impact crater.

Earth certainly encounters meteorites that generate impact craters—scientists have confirmed 190 craters; the action of plate tectonics and weathering has erased all others—but humanity has been lucky over the past few thousand years. The solar system contains several big objects whose orbit might intersect Earth's orbit, but a conjunction has never happened in recorded history. We have, however, had some near misses. In 2013, above the Chelyabinsk region of Russia, a piece of rock speared into the atmosphere at a shallow angle. When it entered the atmosphere it had a mass of about 12,000 tonnes—so it dwarfed the Hoba meteorite. The rock burned up, releasing as it did so about 30 times as much energy as the Hiroshima bomb. The largest fragment to reach the ground weighed just 654 kg. Russia got lucky: no one was killed. Russia was even luckier in 1908, when a meteor burned up over the Tunguska region. The event released the equivalent of 1000 times as much energy as the Hiroshima bomb and it knocked over 80 million trees. Had the collision happened over Moscow the city would have been obliterated.

The dinosaurs were not so lucky: a meteorite ended their 150-million-year-long domination of the planet. The same could happen to us.

3 Waterworld (Map 3)

I am lucky enough to live at the coast and can spend time watching the play of light on ocean waves. My previous home was about as far from the sea as it is possible to be in England. Back then, when I wanted to be near water, I had to make do with a nearby lake. It was a small circular lake, and circumnavigating it at a gentle pace took about 15 min; arithmetic then told me its surface area was about 10 hectares. In an idle moment I once wondered how many lakes of that size, or bigger, existed. The miracle of modern technology means I now know. The HydroLAKES database aims to provide the outlines of all Earth's lakes with a surface area of at least 10 ha. The database contains information on 1,427,688 such lakes.

The map here plots each one of those lakes. The Caspian Sea, the world's largest lake or its smallest sea, depending on your preference, shows up clearly. Most lakes here are mere dots; taken together, though, the dot density is enough to turn much of Canada and Scandinavia blue. Elsewhere they provide a ghostly trace of the world's continents. It is not too much of a stretch to think of Earth as a water world, particularly when you consider that, in addition to the lakes, the majority of Earth's surface is covered by ocean.

But this liquid is not as voluminous as you might think. Collect all Earth's water—from lakes and rivers, ice caps and glaciers, oceans and atmosphere—and you get a sphere about 1385 km in diameter (see the bottom-right of the map for an indication of size). This is enough to supply every person with 72,000 filled, Olympic-sized swimming pools. Most of that water would be undrinkable. A sphere with all the water from just *freshwater* lakes and rivers—the surface-water sources on which *all* life depends—would be about 56 km in diameter, barely noticeable on a map this size. An equal share for each person is 4.8 Olympic-sized pools. Our water supply might seem limitless, but it is a resource we must manage (see Fig. 8).

Our planet got its water when, early in its history, it suffered a bombardment of icy objects not dissimilar in composition to some of the asteroids that occasionally strike Earth today (see Map 2). Earth is now a closed system: we get few fresh deliveries but neither do we do lose much water to space. The amount of *accessible* water, though, could easily shrink where an expanding population has to live in a warming climate. Countries in the Middle East and North Africa are likely to suffer severe water stress in coming decades.

3 Waterworld

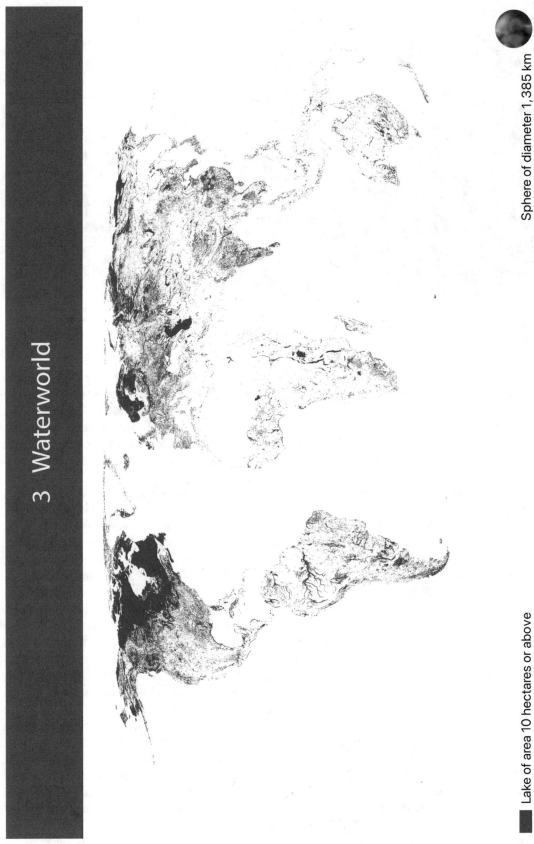

Map 3

Lake of area 10 hectares or above

Sphere of diameter 1,385 km

Fig. 8 Reckless irrigation projects caused the Aral Sea, once the world's fourth largest lake, to shrink. Abandoned ships now litter dusty plains in Uzbekistan and Kazakhstan. This ecosystem collapse led to poor economic and public health outcomes. (Arian Zwegers, CC BY 2.0)

4 Getting Away from It All (Map 4)

The American author Howard Phillips Lovecraft was a complex individual. In life, Lovecraft was a failure: he lived in poverty, often close to starvation; he struggled to work standard hours and was ill prepared to take regular employment; he held virulently racist views, despising anyone who did not share his white, Anglo Saxon, Protestant heritage. His passing at the age of 46, from intestinal cancer, an illness possibly linked to his consumption of canned food that had passed its expiration date, went unnoticed by the world at large. In death, though, Lovecraft became a cultural icon and scholars recognise him as one of the twentieth century's most significant writers of horror.

One of Lovecraft's most important works is his 1928 novella *The Call of Cthulhu*—a story that continues to influence popular culture, from novels by Stephen King through to blockbuster television series such as *Stranger Things*. In *The Call of Cthulhu* Lovecraft provides his readers with the coordinates of a sunken "corpse-city" called R'lyeh, a metropolis "built in measureless eons behind history by the vast, loathsome shapes that seeped down from the dark stars". (Often, Lovecraft's prose was so purple it is better described as ultraviolet.) The city, with a bizarre architecture constructed around the principles of non-Euclidean geometry, forms a prison for an ancient, slobbering monster called Cthulhu. Lovecraft's tale, even today, provides a chill or two.

When Lovecraft began writing this story I presume he gazed at a globe and asked himself: "where can I put my 'corpse-city' so it is remote from any land—the place where people would be most isolated, most distant from rescue?" Anyone looking at a globe and asking that question soon realises the answer must lie somewhere in the Pacific. But, by accident, Lovecraft's location of R'lyeh lies close to the true "oceanic pole of inaccessibility"—the place farthest from terra firma.

Hrvoje Lukatela, a Croatian survey engineer, formally identified the oceanic pole of inaccessibility in 1992, not by sailing there but by using the new generation of computer-based geographical information systems. Geographers subsequently called the place Point Nemo after the character in Verne's *Twenty Thousand Leagues Under the Sea*. The name was certainly appropriate in this context because Nemo is Latin for "No Man". The pole of inaccessibility lies 2688 km away from land. See Fig. 9. After that long stretch of water one can find, more or less equidistant from Point Nemo, three islands: Ducie, one of the Pitcairn Islands; Moto Nui, a small islet belonging to Easter Island; and Maher, off the coast of Antarctica. And none of these islands is exactly abuzz with activity.

The uninhabited Ducie Island, an atoll with an area of just 4 square kilometres, has few visitors. A Portuguese navigator first discovered the atoll in 1606, and in 1790 Edward Edwards rediscovered it while he sailed in pursuit of the mutineers of HMS *Bounty*. The island has only sparse vegetation, but provides a breeding ground for certain seabird species.

Moto Nui, the summit of a mainly submerged volcanic mountain, is little more than a rock. As with Ducie, the land

4   Getting away from it all

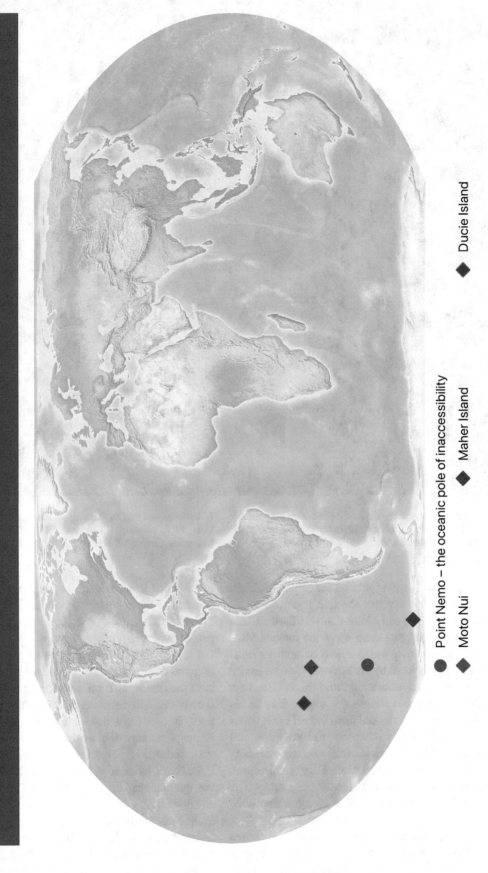

● Point Nemo – the oceanic pole of inaccessibility

◆ Moto Nui   ◆ Maher Island   ◆ Ducie Island

Map 4

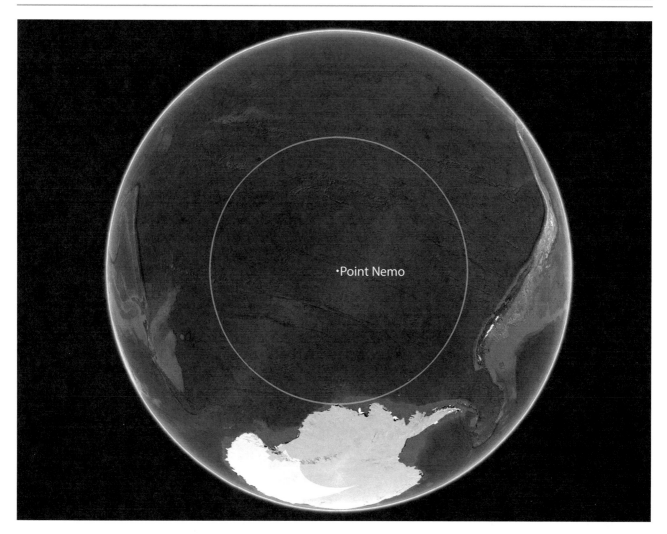

Fig. 9 A view from space, courtesy of Google Earth, above the oceanic pole of inaccessibility. Only a small amount of land is visible: Chile to the east, Antarctica to the south, and New Zealand to the west. The closest lands to Point Nemo—Ducie Island, Moto Nui, and Maher Island—are too tiny to show up at this scale. (Credit: Google Earth; Data SIO, NOAA, US Navy, NGA, GEBCO; Image Landsat/Copernicus; Image US Geological Survey; Image PGC/NASA)

provides an important breeding ground for certain seabirds. The island is also interesting for the role it played in the Easter Island culture. The early inhabitants of Easter Island developed a society that saw them carve the famous Moai statues. That carving society ended due to an environmental catastrophe, and was replaced by the Birdman Cult, an annual competition to determine who would rule Easter Island for that year. Each elder would choose a young man to be his champion. Those youngsters would then dive off a dangerous cliff, swim across the shark-infested waters to Moto Nui, and hide in caves to wait for the arrival of the manutara, or sooty terns. The man who stole the first manutara egg of the season, survived the return swim, and presented the egg intact won the crown for his elder. The practice was not abolished until 1860.

Maher Island was not even discovered until 1946, when the US Navy photographed it while flying missions over Antarctica. The island, which is a horseshoe-shaped piece of rock that is covered in ice for most of the year, gets its name from the commanding officer of the research mission. There is little to see on Maher except for lots and lots of penguins.

One sometimes hears news of an ocean-going vessel getting into difficulty and having to wait for days before another craft can reach it. I am so accustomed to our on-demand society that I find such news difficult to process: how can it possibly take days to reach a ship in distress? But as Point Nemo illustrates, the world remains—in some respects—vast. Indeed, if you ever find yourself floating alone at Point Nemo the nearest people would probably be astronauts: the International Space Station (ISS) orbits at a height of about 400 km, which is much less than the distance between Point Nemo and land.

Even at Point Nemo, though, one can detect a human presence. For example, mention of the ISS brings to mind another link between the pole of inaccessibility and space, a link with observable consequences. Four space agencies (NASA for the USA; ESA for Europe; JAXA for Japan; Roscosmos for Russia) use this part of the ocean, the so-called 'South Pacific Ocean Uninhabited Area', as a cemetery for spacecraft. Much of a spacecraft's body burns up during re-entry, but fuel tanks consisting of carbon-fibre-coated titanium alloys can survive the journey. Since the environs of Point Nemo have the lowest density of humans on the planet and contain the quietest shipping routes on the seas, the space agencies minimise the risk of injury by using this area as a dumping ground for dead craft. The remains of re-entering spacecraft splash and then sink 4 km below the surface. As of 2016, the latest year for which I have full data, parts from 263 spacecraft littered the ocean floor here. (The list comprises 196 Russian objects, 52 US objects, 8 European objects, and 6 Japanese objects; the private SpaceX corporation got in on the act in 2014 when it dumped the second stage of a Falcon 9 rocket.)

A more pernicious form of pollution at Point Nemo involves the plastic we all use and then toss away (see Map 27).

Point Nemo lies within the South Pacific Gyre, a rotating ocean current that blocks cool water from flowing in. The gyre, although it blocks water, does not stop accumulations of plastic waste from forming. The largest patch of waste, about 2500 km north-east of Point Nemo, is a junkyard of polystyrene and cling film, fishing line and plastic pellets. The current first traps the rubbish that humans discard from ships and throw into rivers, then breaks it down into microplastics. Humanity's dross is ubiquitous.

Life at Point Nemo is scarce because sea creatures have little on which to feed: this part of the ocean, being so far from land, receives negligible amounts of windborne organic matter while nutrient-rich water gets blocked by the gyre. The floor beneath Point Nemo is thus one of the least biologically active parts of the ocean.

But even here life finds a way to survive.

At this place two tectonic plates (see Map 7) are pulling apart, creating hydrothermal vents in the gap: cold seawater seeps down into the crust, gets heated by magma, then gushes out of the vents along with a rich stew of minerals. Some researchers believe that life itself might have started in such deep-sea hydrothermal vents. Even if that belief turns out to be wrong, biologists know that some extreme types of bacteria prosper near these vents, thriving on the warm water and the mix of minerals. In turn, yeti crabs—hairy, eyeless crustaceans first identified as recently as 2005—feed on the bacteria.

The blind yeti crab sounds like a creature from Lovecraft's imagination. Could something akin to Lovecraft's Cthulhu also live at Point Nemo?

In 1997, several widely separated underwater listening stations each recorded a strange noise. Oceanologists called it the 'Bloop'. The scientists did not know what caused it, and had only a few pieces of hard information to go on. First, the noise was of a low frequency. Second, for all the microphones to have detected the sound, the Bloop must have been exceptionally loud. Third, it emanated not far from Point Nemo. Fourth, the sound was not made by human technology. Some romantics combined these facts and suggested we had heard the cry of a deep-water leviathan, some sea creature as yet unknown to science. Cthulhu?

No. Not Cthulhu. In the years following the discovery, oceanologists deployed increasing numbers of acoustic sensors close to Antarctica. In 2005, they realised the Bloop was noise from an icequake, a thunderous rumble as an iceberg splintered off an Antarctic glacier. In a way, this was far scarier than Cthulhu or anything else dreamed up by Lovecraft. As our planet warms, the frequency of icequakes will increase. With each Bloop comes the potential for sea levels to rise. And rising sea levels will cause misery for millions (see Map 31).

5   The Longest Straight Line You Can Sail (Map 5)

Patrick Anderson, an American attorney specialising in environmental law, asked himself a question: what is the farthest he could sail in a straight line without hitting land? In 2012 he found an answer and, under his reddit handle of *kepleronlyknows*, posted a map claiming to show the longest straight-line path over water. (A straight line is the shortest path between two points. On a spherical Earth a straight line is an arc of a great circle, which on many map projections—including the one shown here—appears curved.) But was his claim correct?

In 2018, Rohan Chabukswar and Kushal Mukherjee, two computer scientists, published an answer to the question. They developed an algorithm, essentially a set of rules for a program to follow, that enabled a computer to find the required path in a reasonable time. Without their algorithm a computer search would take too long to be practical. Chabukswar and Mukherjee used a method for solving optimisation problems—situations where you want to find the maximum or minimum feature of a system—that dates back to 1960: the "branch-and-bound" technique. The approach works by bunching similar paths into the same branch and then comparing branches. When the algorithm identifies the most successful branch it splits that branch into sub-branches and then repeats the process. The algorithm reduces the number of paths a computer need consider. It took only 10 min for the Chabukswar–Mukherjee program to show that *kepleronlyknows* was right: the longest straight-line path goes from Sonmiani in Pakistan to the Karaginsky District in Russia. See Fig. 10. You will sail 32,089.7 km if

5 The longest straight line you can sail

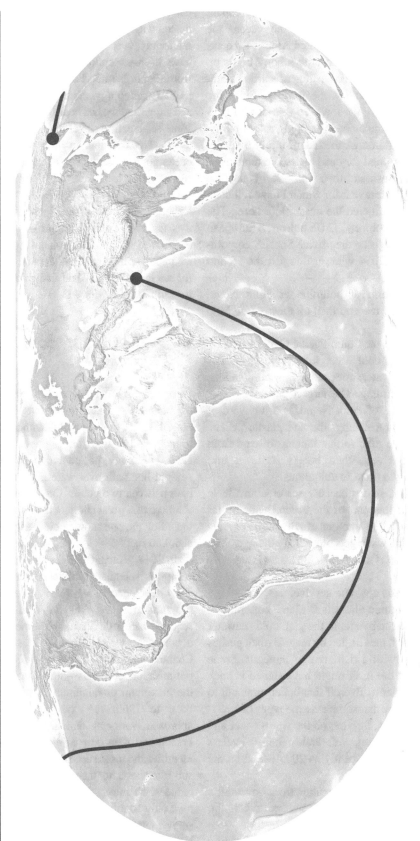

● Sonmiani, Pakistan; Karaginsky District, Russia

— The longest straight line you can sail (which looks curved because of the curvature of the Earth)

Map 5

Fig. 10 Two endpoints of a long straight-line journey. Left: a view of Sonmiani Beach in Pakistan (© Alina Maniar, used with permission). Right: a ribbon seal on the icy Karaginsky Island in the Bering Sea (Vladimir Burkanov (NOAA), CC BY 3.0)

you make the journey. For comparison, Earth's equatorial circumference is 40,075 km.

What would you see if you made such a maritime journey?

Sonmiani, one of the two possible starting points, is a popular beach resort about a two-hour drive from Karachi. (Sonmiani, incidentally, has a port: its *spaceport* offers a launch facility for sounding rockets. See Map 36 for more about spaceports.) From Sonmiani you would sail south, on a route passing between Africa and Madagascar. After skirting the coast of Mozambique and South Africa you would then head into the South Atlantic.

Once in the clear blue you would not see land until you came to the 'Land of Fire', Tierra del Fuego, the archipelago off the southernmost tip off the South American mainland. Charles Darwin made a famous voyage here on HMS Beagle. After rounding the archipelago you would head into the Pacific—and sail a long, long stretch of water without seeing land until you reached your destination: Karaginsky District. This is a small administrative region of Kamchatka Krai, a remote territory of Russia, which holds the distinction of being the site of the highest active volcano (see Map 7) on the Eurasian continent.

To the best of my knowledge, no one has yet made this voyage.

6 The Longest Straight Line You Can Walk Without Getting Your Feet Wet (Map 6)

I have long been a sucker for travel books. My appetite for this genre probably started with the sort of stories I read as a child. Although I knew they were fictional, I longed to visit places such as Asimov's Trantor and Clarke's Rama, Niven's Ringworld and Tolkien's Mordor. Later, when I came across *Jupiter's Travels* by Ted Simon, I realised adventures could happen in real life as well as in the pages of science fiction. Simon set off from London one rainy October day in 1973

and spent the next four years riding 100,000 km on a Triumph Tiger 500 cc. I found something deeply appealing about the idea of a young man deciding to travel around the world on two wheels, despite never having ridden a motorbike before. Simon's book led me to Robert Pirsig's *Zen and the Art of Motorcycle Maintenance*, a story not only of a physical journey but a strangely disquieting spiritual and intellectual trek. I read Paddy Fermor's *A Time of Gifts*, the author's reminiscence of a walk he made from Hook of Holland to Constantinople as a teenager in the 1930s. After that, I devoured accounts of Shackleton's unbelievable journey to South Georgia Island after pack ice crushed his ship *Endurance* (see Map 1). I videotaped Michael Palin's TV series *Around the World in 80 Days*, and imagined emulating his trip. And then... well, after a while, it seemed bookshops and TV listings were crammed with travelogues of every possible type. The magic dulled, somehow. But it made me wonder: is it possible, in this modern era, to devise a journey that has not already been trekked, written about, and then turned into a TV miniseries?

Well, here is an idea: traverse the longest path on land that follows a straight-line while avoiding any major body of water.

The question becomes: what is that longest path? (As we saw in Map 3, there are *lots* of bodies of water to avoid!)

In 2010, Guy Bruneau—a specialist in geographical information systems—triggered an internet discussion about this question when he discovered what he claimed to be the longest coast-to-coast, straight-line path that avoids major bodies of water. Bruneau's path runs from Shitangzhen, China to Greenville, Liberia, and is 13,589.31 km long. If you want to take this route then you need to start in China (because the Chinese authorities let you *leave* China along this route but they will not—at the time of writing—let you *enter* China from this direction). From China you traverse the following countries: Tajikistan, Afghanistan, Tajikistan again, Uzbekistan, Turkmenistan, Iran, Iraq, Jordan, the

6 The longest straight line you can walk without getting your feet wet

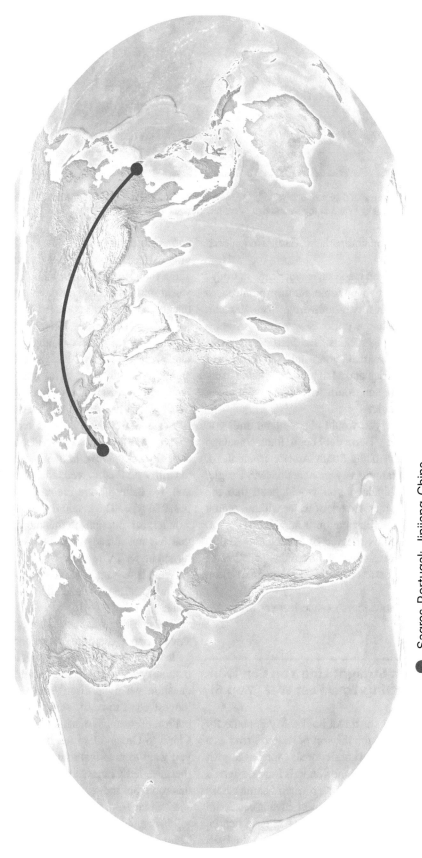

● Sagres, Portugal; Jinjiang, China

— The longest straight line you can walk (which looks curved because of the curvature of the Earth)

Map 6

Fig. 11   Left: Sagres harbour, Portugal (Peter Broster, CC BY 2.0). Right: Hejiang Pavilion in Jinjiang, China (Nyx Ning, CC BY 3.0)

West Bank, Israel, Egypt, Libya, Chad, Niger, Burkina Faso, Ghana, Burkina Faso again, Côte d'Ivoire, and finally Liberia.

There are two problems with Bruneau's suggested path.

First, it is highly unlikely that anyone could attempt the journey in the foreseeable future. The path contains many natural barriers, including mountains, deserts, and extremes of heat. The path also contains artificial hurdles: the political situation within several of those countries, and between neighbouring countries, is dire. The natural barriers are formidable; I suspect the political ones are insurmountable. None of this invalidates Bruneau's claim that Shitangzhen to Greenville is the longest straight-line path avoiding water, of course. But it is disappointing.

Second, and more importantly, I do not believe that Bruneau's path *is* the longest one, because it does not satisfy the constraint he set himself. Follow this route and you go straight through Lake Tharthar in Iraq, through the Dead Sea, and through the Suez Canal. Lake Tharthar is an artificial lake and the Suez Canal is manmade, so perhaps these should not count. But surely the Dead Sea counts as a 'major body of water'?

So what *is* the longest land-based, straight-line path on Earth?

Well, it is exceptionally difficult to check all possible paths. One approach to the problem is to apply the Chabukswar–Mukherjee 'branch and bound' method (see Map 5). As we saw in the earlier discussion, the method compares groups and then sub-groups of similar paths in order to hone in on an optimum solution. It might help to understand the result if we look at the method in a little more detail.

Chabukswar and Mukherjee began by using a map from the National Oceanic and Atmospheric Administration (NOAA), an American environmental agency that aims to understand and predict changes in climate, weather, ocean, and coasts. The NOAA map pictured Earth's surface at a resolution of 1.6 km. In order to simplify their mathematical

model, Chabukswar and Mukherjee agreed to consider anything higher than sea level to be land, and everything below sea level to be water. (Clearly, this assumption does not reflect the messy reality of planet Earth, but the goal of *any* mathematic model is to capture the essentials of a problem while rendering the problem tractable.) Using this mathematical model the branch and bound method was not guaranteed to find the correct solution: the Chabukswar–Mukherjee algorithm was much better suited to identifying a line in water that avoids land than it was to finding a line over land that avoids water. The algorithm in the latter case required several times more computer processing time than the former, and whether it identified the 'correct' path was open to question—adopt a different mathematical model, or apply a more fine-grained data set, then a better path might be identified. There was also the ambiguity, as discussed above, of defining what should be classed as a major body of water. But if nothing else the branch and bound algorithmic approach had the advantage of rigour, and Chabukswar and Mukherjee were able to compare millions of possible paths.

And their result? The longest straight-line path over land, avoiding major bodies of water, runs from near Sagres on the south west corner of the Algarve in Portugal through to a place near the city of Jinjiang in China. See Fig. 11.

The Chabukswar–Mukherjee path is 1831 km shorter than Bruneau's, but it certainly avoids any 'major body of water' as large as the Dead Sea. As in Map 5, the Chabukswar–Mukherjee path shown here looks curved. But, as before, the apparent curvature is because Earth's surface is curved. A Google Earth image, which pictures how a road between these points might look from space, demonstrates the path is straight. See Fig. 12.

Despite the title of this section, you *would* have to roll up your trousers for this journey: fire up Google Earth and you will see the path cuts across a few ponds and a number of streams (the mathematical model that Chabukswar and Mukherjee developed explicitly excluded rivers as a

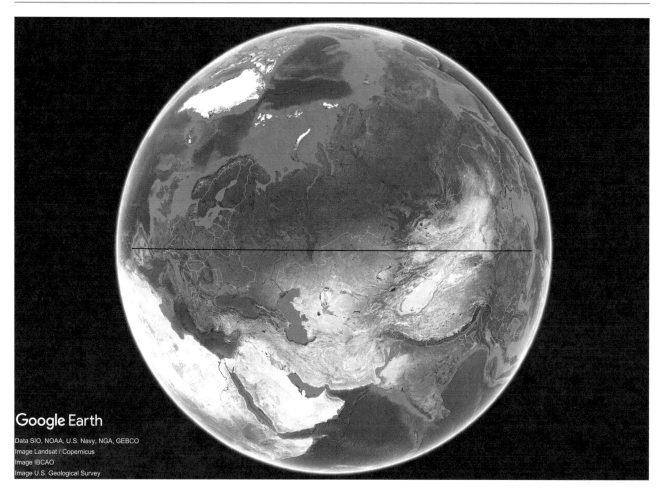

Fig. 12 A view from space, altitude 12,000 km, of the path from Sagres, Portugal to JinJiang, China. As is clear from this image from Google Earth, the path is straight and it avoids any major bodies of water—the two constraints we have set ourselves. (Credit: Google Earth; Data SIO, NOAA, US Navy, NGA, GEBCO; Image Landsat/Copernicus; Image IBCAO; Image US Geological Survey)

variable). But the journey would be remarkably dry. Suppose your starting point was Sagres, at the extreme southwestern tip of Europe. You would find yourself in a small parish that enjoys some of the sunniest weather in the continent. The path would then take you through 15 countries as you headed east: Portugal, Spain, France, Switzerland, Liechtenstein, Austria, Germany, Czech Republic, Poland, Ukraine, Belarus, Russia, Kazakhstan, Mongolia (very briefly, before returning to Kazakhstan), and finally China. Your end point, Jinjiang, is a bustling city of two million.

To my knowledge, no one has trodden this path or even attempted to tread it. If you did try it, you would find much of the middle part of the journey—a path through Poland, Ukraine, Belarus, and Russia—to be relatively flat. You would encounter difficulties such as forests, of course, and some of the terrain would be challenging for a number of reasons, but elevation would not pose too much of a problem. You might need your climbing boots at the start of the journey, though, since you would need to get across the Pyrenees. And you would encounter mountainous regions

towards the end of the journey, as you moved through Kazakhstan and Mongolia into China. I am sure such a journey would have been beyond me when I was a young man; now I am an old man it would be impossible. Nevertheless, the thought of attempting an expedition that no-one has tried before is a romantic one. If any TV producer reads this: I'm willing to offer my services for a Palin-style adventure!

7 Volcanic Eruptions Over the Past 12,000 Years (Map 7)

The Smithsonian Institution's Global Volcanism Program maintains a database of the last known eruption of all volcanoes throughout the Holocene (the name of the current geological epoch, which began just under 12,000 years ago). At the time of writing the database contains details of 1357 volcanic eruptions and it makes for fascinating reading.

The oldest volcanic activity in the database relates to Mount Nantai on Honshu, Japan's main island. Nantai is a

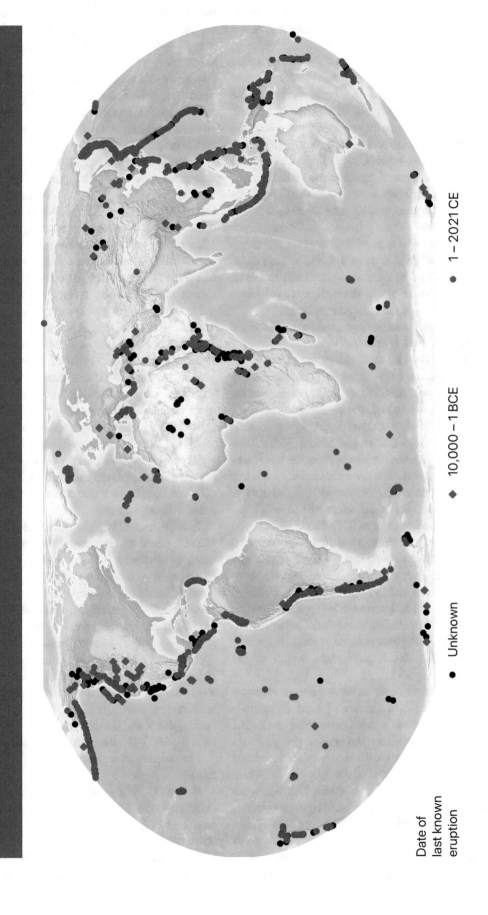

7   Volcanic eruptions over the past 12,000 years

Date of last known eruption

• Unknown

♦ 10,000 – 1 BCE

• 1 – 2021 CE

Map 7

stratovolcano. (A stratovolcano is the commonest form of volcano, and consists of many layers of hardened lava and other material laid down after repeated eruptions. The layers build up to generate a mountain with a steep profile with a crater at the summit. When we hear the word 'volcano', most of us will conjure up a picture of a stratovolcano in our mind's eye.) The last eruption of Nantai dates back to 9540 BCE. Presumably some humans will have seen the eruption, but of course no records from that period have survived. The oldest volcanic eruption for which there might be recorded evidence is, according to the database, that of Yucamane in Peru: this stratovolcano last erupted in 1320 BCE. The most recent eruption in the database is that of La Palma, one of Spain's Canary Islands; as I write, television news channels have been providing some spectacular—and scary—footage of lava flows on La Palma. That information will be out of date by the time you read this, however, because at any given time typically twenty or more volcanoes are erupting somewhere. Earth is an active planet.

As a glance at Map 7 clearly shows, the volcanoes in the Smithsonian's database are not randomly distributed over the surface of the Earth. The United Kingdom, for example, has experienced no volcanic activity during the Holocene. (A city such as Edinburgh owes much of its current grandeur to the aftereffects of a volcano; indeed, the particular form of the Arthur's Seat volcano even shaped Edinburgh's political history. But that volcano last erupted about 340 million years ago.) A short distance northwest from the UK, however, brings you to an island whose map representation is lost under the dots of Holocene eruptions: Iceland is home to 30 active volcanoes, of which just under half have erupted since the Norsemen settled there in 874 CE. (In Iceland you can see shield volcanoes, which are different to the usual stratovolcanoes. A shield volcano forms when low viscosity lava spreads in a thin layer over the surrounding area, building a low-profile feature said to resemble a warrior's shield lying flat on the ground.) A fact even more apparent from the map is the existence of volcanic 'chains': the eastern hemisphere contains some interesting patterns of volcanic activity, while one chain runs down the western length of the Americas.

This clustering of volcanic activity happens because Earth's surface consists of a number of rigid segments—tectonic plates—floating on a soft layer of hot rock. The plates move slowly, just a few centimetres per year, but inexorably. When two plates collide, one plate is pushed beneath another. The plate that sinks gets hotter, which causes water to be released from minerals. This hot water rises up and eventually, in a complicated process, causes rocks just below the surface to melt. Where this molten rock manages to rise and erupt on the surface we see a volcano. When two plates move apart, on the other hand, magma rises straight up and erupts on the surface as lava. So

the location of plate boundaries explains the location of volcanoes. The Pacific Plate, for example, forms the bed of the Pacific Ocean; and this oceanic plate finds itself surrounded by various continental plates, including the Australian–Indian plate, the Eurasian plate, the North and South American plates, and the Philippine plate. The relative motion of the Pacific oceanic plate and the surrounding continental plates causes the 'Ring of Fire' so clearly visible on the map.

A few volcanoes are located in the middle of a plate. These volcanoes occur when a hot plume rises up from deep in the mantle, carrying magma to the surface. The islands of Hawaii, for example, are believed to have been created from a hotspot. Such volcanoes are no less impressive. Mauna Kea, one of the five volcanos that make up the Big Island of Hawaii, is the tallest volcano in the world. Measured from the sea floor, Mauna Kea is more than 9 km meters tall: the tallest mountain on Earth. Its summit is 4.205 km above sea level, which is one of the reasons why astronomers use Mauna Kea as a base for telescopes.

Volcanoes can, of course, be dangerous. According to a researcher at the University of Bristol, about 800 million people live within 100 km of an active volcano—and thus within the range of what can sometimes be lethal hazards. Most victims of volcanic activity have fallen victim to pyroclastic flows (fast-moving avalanches of hot ash, gas, and rock) and lahars (mudflows created from a mixture of rainwater or meltwater and volcanic ash), though people can also die from inhaling toxic gas or, if their escape routes are cut off, from lava. But a volcano can affect people far beyond its immediate environment.

In 2010, a small volcano with a colourful name—Eyjafjallajökull; see Fig. 13—caused the closure of airspace over much of northwest Europe. Iceland is the site of several much larger volcanoes than Eyjafjallajökull, but this particular eruption caused havoc because of the amount of ash it injected into the jet stream. Volcanic ash is a flight safety hazard. Its hard, abrasive nature can damage propellers and scour windows; ash can contaminate a plane's water and fuel systems; in some cases it can cause engine failure. When Eyjafjallajökull erupted, therefore, many European countries were forced to close their airspace to commercial jet aircraft. Millions of passengers had their travel plans disrupted (among them my German mother-in-law, who was visiting us in the UK.) The disruption lasted for a week.

The impact of volcanoes can be traced back through history.

In 1883, four violent volcanic explosions almost obliterated the island of Krakatoa. The third explosion generated the loudest sound ever recorded on Earth: instruments detected the sound wave through three and a half circumnavigations of the world. The eruption killed tens of thousands of people, and had a planetary-wide impact:

Fig. 13 The stratovolcano Eyjafjallajökull has a height of 1.65 km; at the top is an open crater. In April 2010, Eyjafjallajökull erupted. Although small, the explosion was sufficient to disrupt air travel in Europe for a week. Prior to that, the last major eruption of the volcano happened in the nineteenth century: activity began in December 1821 and a series of eruptions then followed, ending a year later in January 1823. Ash from these eruptions can still be seen on Iceland. Considering the duration of this nineteenth century event, the 2010 eruption was nowhere near as disruptive as it might have been. (Bjarki Sigursveinsson, CC BY-SA 3.0)

the ash, thrown high into the atmosphere, lowered global temperatures for five years.

A few decades earlier, the 1815 explosion of Mount Tambora killed more than 70,000 people. Volcanic ash, thrown into the atmosphere, disrupted weather patterns: 1816 became known as the year without a summer, the northern hemisphere's second-coldest year in six centuries. (The dreary weather caused Mary Godwin, Percy Shelley, John Polidori, and Lord Byron to sit around a log fire and make up ghost stories. We have Tambora to thank for *Frankenstein*.) The eruption led to poor harvests in northern Europe, the death of livestock in America, and an increase in the severity of a typhus epidemic in southern Europe.

Further back in time, pyroclastic flows from the eruption of Mount Vesuvius in 79 CE buried the Roman cities of Pompeii and Herculaneum. We can still see the preserved bodies of those unfortunate people who were unable to escape. And the eruption of Thera in around the sixteenth century BCE devastated the beautiful island that is now called Santorini; perhaps the event inspired the Greek myth of Atlantis.

Even further back in time, 75,000 years ago, a supervolcano occurred in what is now Lake Toba in Sumatra. A supervolcano is so violent it has the potential to trigger a mini Ice Age and cause species to go extinct. Fortunately for humans, we have not experienced a supervolcano in recorded history. The Toba supervolcano was vast. There is a suggestion—a plausible suggestion, but one lacking conclusive evidence to support it—that Toba triggered a decade long global winter that drove humans to the brink of extinction. Some studies indicate the presence of a 'genetic bottleneck' 70,000 years ago; the human population might have dipped to as few as 3000 people. The theory remains controversial, but the fact it can be discussed at all demonstrates the power of volcanoes.

Volcanoes have an importance beyond their toll on human life and their effect on human culture. It is possible that all life on Earth owes something to volcanoes.

Some geophysicists argue in favour of the so-called Snowball Earth hypothesis: the notion that, at least once in its history, Earth froze. In a Snowball Earth the whole globe might have resembled present-day Antarctica. This sounds disastrous for life, but a global glaciation event might have provided the conditions under which unicellular organisms could give rise to multicellularity. If that is the case then complex life is here because of a Snowball Earth. But there is a problem with the hypothesis: it is easy to imagine a scenario whereby Earth becomes covered in ice, more difficult to imagine how the ice later melted. That is where volcanoes come in. Volcanoes belch CO_2. Normally, the weathering of rocks withdraws this greenhouse gas from the atmosphere—but this could not happen on an ice-covered Snowball Earth. Over millions of years, the CO_2 emitted by volcanoes would build up. Volcanoes caused a much-needed dose of global warming.

8   Forests (Map 8)

Global Forest Watch, an interactive online platform, allows people to monitor forests using more or less real-time data. (Incidentally, according to the UN's Food and Agriculture Organization (FAO) the definition of a forested area is "land under natural or planted stands of trees of at least 5 m *in situ*, whether productive or not" and the definition "excludes tree stands in agricultural production systems".) The platform tells us, for example, that forest covers roughly 31% of Earth's land surface. That equates to about 4 billion hectares of forest (which, for those of you who prefer to work in different units, is 40 million square kilometres or about 15.6 million square miles). For comparison, the combined area of North and South America is 4.255 billion hectares. So Earth has a surprising amount of forest cover—certainly more than I thought was the case. We are, however, losing trees. According to the FAO, Earth had about 5.9 billion hectares of forest before the Industrial Revolution. Over the past couple of centuries, humans have systematically removed roughly a third of the planet's forest cover.

If human civilisation is to have a long term presence on Earth then surely we need to recognise that this systematic deforestation is one of the more dangerous activities in which our species is engaged. Forests are important for a number of reasons.

Forests, along with phytoplankton, are the lungs of our planet. They absorb carbon dioxide (the stuff we exhale) and produce oxygen (the stuff we inhale). They help us breathe.

Tropical forests store about a quarter of a trillion tons of carbon. Only the oceans form a bigger carbon sink. By now few people can be unaware that the release of greenhouse gases such as carbon dioxide is fueling the climate change emergency.

Forests are places of biodiversity. Biologists estimate that about half of all known species live in forests. When we cut down forests we lose more than just trees; we lose mammals, birds, and insects. When we displace creatures from their natural habitat, and force them into closer contact with humans, we increase the risk of pandemic.

We are currently undergoing the sixth great mass extinction of species. Human activity is driving extinction at a rate 1,000 to 10,000 times beyond natural levels. Protecting forest habitats is key to protecting our planet's remaining biodiversity.—Global Forest Watch (n.d.)

If a forest is on a sufficiently grand scale it can influence regional weather patterns. The Amazon rainforest, for example, affects local rainfall—but also rainfall in places as distant as the Great Plains in the USA. As well as promoting rainfall, forests fight flooding: tree roots, for example, help with absorption, and they slow the flow of flash floods that might otherwise wash away soil.

People have long known that certain plants have medicinal effects. The rainforest, in particular, is teeming with potentially therapeutic agents: native plants have adapted to a crowded environment by developing a range of complex chemical responses to the threats posed by the insects and pathogens that thrive in these warm, wet conditions. When we destroy forests we risk missing out on new drugs.

Forests provide jobs and homes for many millions of people, and they provide wellbeing and enjoyment for millions more.

Forests are important. And yet we continue to destroy them.

In 2018, according to data from Earth-monitoring satellites, human activity destroyed 12 million hectares of tropical forest. Instead of hectares or square kilometres or square miles, let's use the traditional journalistic unit for expressing area: the football pitch. In these units, the rate of tropical forest loss was half a football pitch *every second*. The loss was less than in the years 2016 and 2017, but only because those two years were particularly dry—millions of hectares went up in flames. The Covid-19 pandemic acted as a break on many activities, of course, but the trend of deforestation is upwards, with all the threat to runaway climate change and species mass extinction this implies. So why are we doing it? In recent years the main drivers of deforestation have been people's appetite for beef, chocolate, and palm oil.

In Brazil, cattle ranchers have destroyed vast swathes of rainforest (see Fig. 14). They are willing to drive deep into the indigenous territories that are home to some of the world's

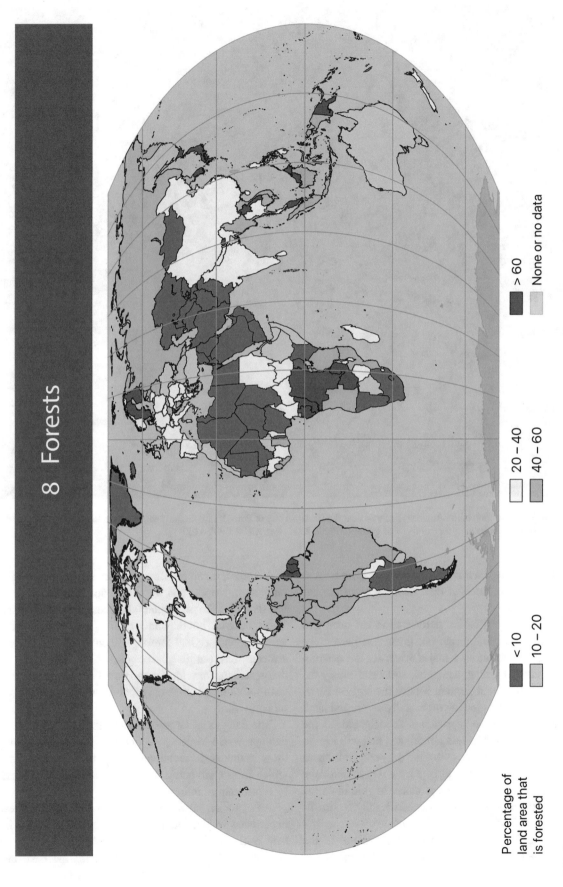

8 Forests

Percentage of
land area that
is forested

■ <10
□ 10 – 20
□ 20 – 40
■ 40 – 60
■ >60
□ None or no data

Map 8

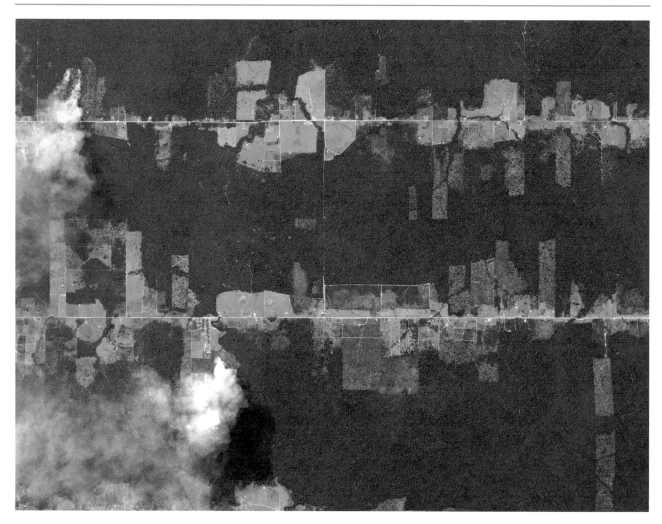

Fig. 14 Systematic destruction of the Amazon rainforest near the Brazilian city of Porto Velho, in the upper Amazon River basin. Fires are used to clear the rainforest so it can be used for farming and grazing. (Planet Labs, Inc., CC BY-SA 4.0)

last remaining uncontacted tribes. The ranchers rip up trees to create grazing pasture; the beef then finds its way onto our supermarket shelves.

In Ghana and Côte d'Ivoire, deforestation has been less than in Brazil in absolute terms but greater in percentage terms. In these two countries a major driver for the destruction has been cocoa farming, which takes place to satisfy the West's sweet tooth. Local farmers believe the biggest beans come from cocoa plants growing on recently deforested land. Even if that were true the approach is shortsighted: as the forest vanishes there will be less rainfall and that will lead to worse crops. Cocoa, however, is not the only culprit in these countries: illegal gold mining also plays a significant role. Ghana, for example, is the biggest producer of gold in Africa but about one third of the total is mined illegally by small-scale operators.

In Indonesia, forests have been targeted because of the West's use of palm oil. In addition to palm oil's application

as a biofuel, it can be found in biscuits, candles, detergents, margarine, soap... and we ensure our supplies by replacing forests with palm oil plantations. These plantations are called 'green deserts', but they have nothing approaching the biodiversity of the original forests. Although the Indonesian government has put in place policies to slow the pace of deforestation, for which it should be congratulated, a longer term solution will come only if civilisation lessens its reliance on palm oil.

The dangers of large-scale deforestation are so clear that perhaps world leaders will one day follow the example of those in Indonesia, and implement policies to slow the rate at which forests are being denuded—followed, we must hope, by a program of reforestation.

As of today, however, where can you go if you want to experience pristine, verdant forest? On that score it's hard to beat Suriname, a small country with Brazil to the south, Guyana to the west, French Guiana to the east, and the

Fig. 15 Suriname has the highest percentage of forested land of any country. This view of the Suriname river, looking down from Blue Mountain, is a typical sight here: deforestation has not occurred on the same industrial scale as in neighbouring Brazil. (JvL, CC BY 2.0)

Atlantic Ocean to the north. Most of Suriname's population of just over half a million lives on the coast. Much of the land is unspoiled forest; see Fig. 15. It is home to many unique species, the blue poison dart frog being the most famous. The Central Suriname Nature Reserve, which covers 16,000 square kilometres, is a UNESCO World Heritage Site (see Map 33) because of its forest and the biodiversity it supports. As is the case with forests everywhere, however, its future might come under threat. Suriname has large deposits of gold and, as prices rise, so do the financial rewards of mining.

And if you hate forest? No trees or shrubs grow in the wastes of Antarctica, of course. Not surprisingly, some small countries, such as Monaco, possess no forests. More surprising is Qatar. The lack of natural forest isn't the surprise: this is a desert country. The surprise, rather, is the creation of an artificial forest in the desert. The Um Salaal forest project has planted more than 100,000 trees, mainly hardwood trees that don't need large amounts of water to survive, around two artificial lakes. Irrigation depends on treated water from a nearby sewage treatment plant. If the project succeeds, the inhabitants of Qatar will certainly benefit.

9   Chemical Elements Named After Places on Earth (Map 9)

The substance of the world seems almost impossibly varied, dizzyingly rich. Over the centuries, however, chemists have uncovered hidden order: only a limited number of fundamentally different substances exist. A *chemical element*, or just

plain *element*, is a substance that cannot be decomposed or changed into another substance using chemical means. You can think of an element as being a basic chemical building block of matter. The varied and rich substance of the world arises from combinations of that limited number of elements. Appreciable amounts of 90 elements occur naturally, most of which were baked inside a star that exploded and seeded the protoplanetary disk around the young Sun (see Map 1); depending upon whom you ask, another 4 or 8 elements occur naturally due to the radioactive decay of heavier elements. At the time of writing chemists know of 118 elements; 20 of those elements have been synthesised only in the laboratory.

(As an aside: an element is a substance containing just one type of atom. In turn an atom is characterised by the number of protons carried in the atomic nucleus. Hydrogen, for example, has one proton in its atomic nucleus; helium has two; lithium three; and so on up to oganesson, which has 118 protons in its nucleus. Since chemical reactions do not affect the atomic nucleus, elements cannot be changed chemically. Physicists, however, can initiate *nuclear* reactions—and thus realise the ancient dream of alchemists: transmutation. Physicists have even learned that the particles making up an atomic nucleus—protons and neutrons—are themselves composite. But let's return to the topic at hand.)

People discovered the chemical elements in different places and at different times (some were known in antiquity; oganesson was first synthesised in 2002). The names of the elements, therefore, derive from different languages and cultures. A total of 32 elements derive their (English) names from places on Earth. The map here shows the inspiration for the names of these 32 elements.

9 Chemical elements named after places on Earth

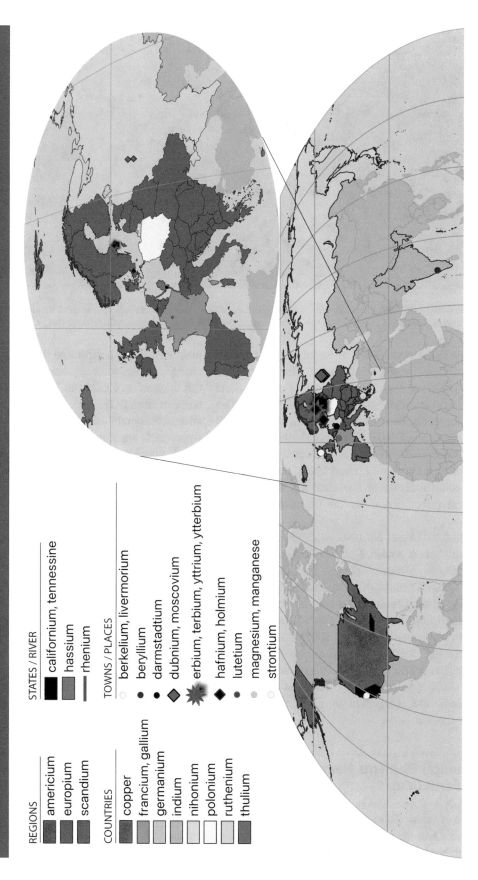

REGIONS
- americium
- europium
- scandium

COUNTRIES
- copper
- francium, gallium
- germanium
- indium
- nihonium
- polonium
- ruthenium
- thulium

STATES / RIVER
- californium, tennessine
- hassium
- rhenium

TOWNS / PLACES
- berkelium, livermorium
- beryllium
- darmstadtium
- dubnium, moscovium
- erbium, terbium, yttrium, ytterbium
- hafnium, holmium
- lutetium
- magnesium, manganese
- strontium

Map 9

Fig. 16 Waste rock outside the Ytterby mine in Sweden. Miners dug here for quartz in the 1500s, and for feldspar starting in 1790. Mineral samples from the mine directly gave rise to the names of four chemical elements, and the names of a further four elements have an indirect connection to the quarry. (Bengt Oberger, CC BY-SA 4.0)

One could look at those 32 elements in a variety of ways: by their location in the periodic table; by their number of protons; by their properties. Let's consider them by order of discovery.

Copper, then, comes first. Copper was the first metal humans learned to mine and craft—people discovered the substance about 11,000 years ago and by 5000 BCE smiths were smelting copper. The English name for the metal comes from the Latin *cyprium*, which means 'from Cyprus'—the island that produced so much of the material in ancient times. The English simplified the Latin *cyprium* to the anglicised *cuprum*, which gives us the symbol for copper: Cu.

Millennia later, in 1755, the Scottish chemist Joseph Black observed the element magnesium; Humphrey Davy isolated it in 1808. The name comes from Magnesia, a district in the Greek region of Thessaly, from where minerals containing the element were mined. The same place also gave its name to manganese, discovered in 1774.

Strontium, a soft metallic element with a yellowish hue, takes its name from Strontian—a hamlet in the Scottish highlands. A mineral containing the element was discovered near the hamlet in 1790; in 1808, the tireless Humphrey Davy isolated the metal.

Yttrium has modern uses in electrical technologies but the element was discovered back in 1794, in a mineral found near the Swedish village of Ytterby. See Fig. 16. At this point I break the strictly chronological ordering to note that Ytterby gives its name to three other elements: erbium, terbium, and ytterbium. (The chemist who discovered yttrium, Johan Gadolin, also gives his name to an element: gadolinium. Ytterby deserves the star I give it on the map.)

Returning to a chronological approach, the next element on our list is beryllium. In 1798, the French chemist Louis Nicolas Vauquelin discovered the compound beryllia in the semiprecious stone beryl. The name of the gemstone, and thus indirectly the element, comes from the Indian city Belur. Again breaking briefly with chronology, India itself indirectly gives its name to an element, indium, discovered by German chemists in 1867. When heated, it emits a bright indigo-blue line in its spectrum. Indigo essentially means 'blue dye from India'.

Fig. 17 Equipment in the GSI Helmholtz Centre for Heavy Ion Research, Darmstadt, Germany. The search for new elements, which have only a fleeting existence, requires complicated instrumentation such as this. (Alexander Blecher, blecher.info, CC BY-SA 3.0)

In 1844, Karl Ernst Claus discovered a new element in samples sent from Russia. He named the element ruthenium, from a Latin word for Russia. In 1875, the French chemist Paul-Émile Lecoq de Boisbaudran found an element predicted by Mendeleev, following the latter's arrangement of known elements into a periodic table. The Frenchman named it gallium, after the Roman name for his native country, but he perhaps also displayed a degree of self-aggrandisement: his middle name Lecoq means 'rooster'—which in Latin is *gallus*.

In 1879, Lars Fredrik Nilson took a sample of ytterbium oxide and found it contained a new element, which he named scandium in honour of Scandinavia. In the same year, Per Teodor Cleve took a sample of erbium oxide and isolated two new elements: holmium and thulium. Cleve named the former after Stockholm, the latter based on a misunderstanding. In ancient times, Thule was a mythical northern land several day's sail north of Britain. By medieval times, Thule was often identified with Iceland (as I have shown it here on the map). Cleve, though, mistakenly thought that Thule was the ancient name of Scandinavia. Despite Cleve's misapprehension, we can see that the mines around Ytterby gave rise, directly or indirectly, to the names of a not insignificant percentage of the elements.

In 1886, the German chemist Clemens Alexander Winkler found another element whose existence Mendeleev had predicted. He named it germanium after his native country. Two further elements were observed before the turn of the century: europium (after Europe) and polonium (after Poland). The latter was observed and isolated by Pierre and Marie Curie.

In 1906, a sample from Ytterby threw up yet another element, lutetium (from the Latin for 'place of mud': Paris). In 1908, the Rhine inspired the name of the element rhenium. The final stable element, hafnium, found in 1922, got its name from the Latin for Copenhagen.

In 1939, Marguerite Perey, one of Marie Curie's students, discovered francium and thus France was honoured a second time. Three elements were then named for the New World: americium (1944), berkelium (1949), and californium (1950) were discovered by scientists at the University of California, Berkeley. In 1970, the Joint Institute for Nuclear Research (JINR), based in Dubna, Moscow Oblast, discovered dubnium. More recently, researchers at the GSI Centre for Heavy Ion Research (see Fig. 17) in the German town of Darmstadt, which is in the state of Hesse, discovered four elements they named after famous scientists (bohrium, meitnerium, roentgenium, and copernicium). And they discovered two elements they named after their locality: hassium (found in 1984) and darmstadtium (1994).

A laboratory in Livermore, California, has collaborated with JINR to investigate extremely heavy elements.

Livermorium (found in 2000) and moscovium (2003) celebrate Livermore and Moscow. In 2003 the US–Russian collaboration found evidence for element-113; around the same time a Japanese group found evidence for the same element. After years of discussion, priority for the discovery was assigned to the Japanese group, who named the element nihonium after the native name for Japan. Finally, JINR collaborated with the Oak Ridge National Laboratory, Tennessee to look for element-117. In 2009, they synthesised tennessine, the most recent element (at the time of writing) to have been discovered.

10   Earth's Phosphorus (Map 10)

All life on Earth depends on six elements: carbon, hydrogen, oxygen, nitrogen, phosphorus, and sulphur. In the absence of these essential substances, life as we know it is impossible. The importance of oxygen, say, is clear. But why is phosphorus necessary? Well, it fulfils a number of biological roles but two functions in particular are fundamental. First, the nucleic acids DNA and RNA—large molecules that play a key role in biology—possess a structural framework that depends on a phosphate backbone. Second, any cellular process that uses energy employs a complex molecule called adenosine triphosphate (ATP) as an energy transporter. No phosphorus, no DNA, no RNA, no ATP. No phosphorus, no life.

One might expect phosphorus, given its importance in biology, to be cosmically abundant. It is not. In terms of cosmic abundance it is a rather scarce element. Indeed, we do not know for certain where Earth obtained its phosphorus. The generally accepted picture is that Earth had already formed and gained its life-giving water before being peppered by phosphorus-containing meteorites. Meteoritic bombardment thus not only has the potential to destroy life, as we saw in Map 2, but to permit life to begin in the first place. (Phosphorus is ultimately cooked up inside stars and distributed through the galaxy by supernovae. One recent piece of astronomical research provokes the tentative suggestion that, in order to possess phosphorus and thus the possibility of life, a newborn planet must form in the vicinity of a particular type of phosphorus-supplying supernova. Might the infrequency with which this happens partly explain why we have yet to observe life anywhere else in the cosmos?)

Regardless of how Earth came by its phosphorus, the total amount of the element is now fixed: scientists cannot fabricate the stuff in a laboratory. Instead, phosphorus molecules take part in an eons-long natural cycle that has a number of subcycles. The element is bound in rock deposits and diluted in water. Plants draw up phosphorus from the ground; herbivores get their phosphorus by eating plants; and carnivores obtain *their* phosphorus by eating herbivores. A typical person ingests about 1 g of phosphorus a day (and stores about 750 g of phosphorus in bones and teeth). Most of the phosphorus creatures ingest is excreted and, at least in the past, farmers returned this vital element to the soil—and thus helped mitigate problems of phosphorus deficiency; see Fig. 18—by spreading manure and natural waste.

And then we went and broke the cycle.

The 1960s saw the most successful revolution of all time: the Green Revolution. Scientists developed synthetic fertilisers, thus expanding the amount of land suitable for agriculture and allowing millions of people to be fed who would otherwise have starved. The revolution was made possible not only through Norman Borlaug's development of better plant varieties but also by the Haber–Bosch process, a technique for obtaining nitrogen from air in a way that makes it available for crops in soil. Nitrogen is one of those six vital elements. But if nitrogen levels increase in soils there needs to be a corresponding increase in phosphorus. That creates a problem because there is only so much phosphorus-rich manure to go round and it is not possible to obtain phosphorus from the atmosphere in the same way one obtains nitrogen. Companies therefore began to get phosphorus in the only way they knew how: they mined phosphate from geologic deposits.

For a while, all went well. The Green Revolution has allowed the population to exceed 7 billion and, in principle, enough food exists to feed everyone. But when farmers and food producers spread synthetic fertilisers on their land, excess phosphorus gets washed away and eventually ends up in lakes, rivers, and oceans. It does not disappear, but it becomes uneconomic to recover. (The excess also causes a form of water pollution called eutrophication, in which algae flourish and then die, depleting oxygen levels and creating a 'dead zone'.) Food production today means a daily human intake of about 1 g of phosphorus results in the depletion of about 22.5 kg of phosphate rock.

You can probably see the problem. When there were not so many mouths to feed, and food producers recycled phosphorus, there was no issue. But to feed billions of people, particularly at a time when the recycling of phosphorus has become less efficient, we need to mine large quantities of phosphate rock. And quantities of *any* geological resource—coal or oil or phosphate—are necessarily limited. So the question becomes: might we run out of phosphorus?

> There are no substitutes for phosphorus in agriculture.—US Geological Survey (2020)

Journalists are wont to declare, every few years or so, that humanity will soon hit 'peak phosphorus' and the phosphate rock reserves from which we can economically extract phosphorus will soon be gone. This is unlikely. Back in 2010, the

10 Earth's phosphorus

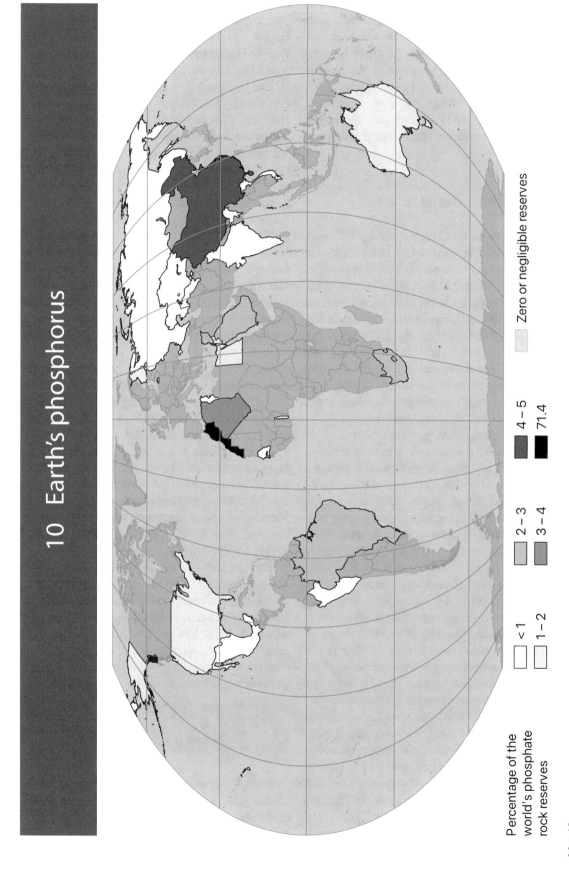

Percentage of the
world's phosphate
rock reserves

<1	4 – 5
1 – 2	71.4
2 – 3	Zero or negligible reserves
3 – 4	

Map 10

Fig. 18 Maize growing in acidic soils on a farm in KwaZulu-Natal, South Africa. The soil in the background has been limed and the plants grow strongly. Plants in the foreground suffer from aluminium toxicity due to the soil's acidity. Poor root growth causes a number of problems, including phosphorus deficiency; in turn, this causes small leaf size and can prevent shoot growth. (Alandmanson, CC BY 4.0)

US Geological Survey estimated the global reserves of phosphate rock to be just under 18 billion tonnes; in 2019, the same organisation estimated there to be 70 billion tonnes of phosphate rock. Our supply should last us for several centuries. If we *were* to run out of phosphorus that would indeed spell trouble because we have no alternative substances we could use. But we have ample reserves. The reason we need to take the issue of phosphorus security more seriously revolves around where the reserves are concentrated. Just look at the map.

More than 70% of the world's phosphate rock reserves is found in Morocco and Western Sahara. The longest conveyor belt ever built takes white phosphate rock from desert mines to the port of El Aaiún, and then ships transport the rock all around the globe. See Fig. 19. But this region is a place of conflict. Western Sahara is disputed land.

The Sahrawi people have lived in the region since before colonial times, and have been fighting for self-determination since 1975 when the colonial power (Spain) allowed Morocco and Mauritania to split the area. The UN refers to Western Sahara as a non-self-governing territory. Morocco actually controls most of the territory as Moroccan Sahara, but no other country recognizes the legality of its claim. To complicate matters further, the Algerian-backed Polisario Front have engaged in a military dispute to gain independence for the territory as the Sahrawi Arab Democratic Republic. Without getting into the rights-and-wrongs of a complex situation, one can still point out that much of the world's supply of a critical resource lies in the middle of a region of significant political tension. It is not difficult to imagine situations in which the price of phosphate rock might skyrocket.

This happened in 2008: an increased demand for fertiliser (caused by a global increase in meat consumption), combined with higher oil prices and a short-term squeeze on the availability of phosphate rock, led to a 800% increase in the price of phosphate. Food prices surged, which hit developing countries with particular force.

And *this* is the problem. The world has large reserves of phosphate, but our management of this vital resource is poor. There will soon be 9 billion people on this planet, an increasing fraction of whom appear to want to eat meat and diary; farmers and agriculturalists will have to battle against the ravages of climate change; the quality of phosphate rock reserves will decline and require more energy to mine. All these factors require us to take the management of phosphorus more seriously. There is no reason why we are *unable* to do this. But doing so requires political will.

Fig. 19 This photograph, taken by the astronaut Alexander Gerst, shows phosphate mining in the desert of Western Sahara. The near-vertical straight line on the left-hand side of the photograph is the start of the world's longest conveyor belt, which is used to take phosphate rock to port for shipping around the world. (ESA/A.Gerst, CC BY-SA 3.0)

The World of Countries

In the first chapter we looked at the natural world which, oblivious as it is to artificial boundaries, does not recognise national borders. If we needed proof of this fact, the spread of the SARS-CoV-2 virus provided a demonstration. In search of hosts the expanded its reach, not pausing at frontier checkpoints or border control posts; soon, it was everywhere. But if you choose to look through the lens of nation states then the world begins to exhibit differences. Consider, for example, the continent of South America.

South America contains twelve sovereign states, ranging in size from giant Brazil (the continent's largest and most populous country) down to tiny Suriname (South America's smallest and least populous country). Also on the continent is French Guiana. This is roughly half the size of Suriname in terms of area and population. So why does French Guiana not count as the continent's smallest country? Because it is an overseas department of France. Although French Guiana bears no relationship to France in terms of its natural characteristics, in terms of its political identity it *is* France. On any map that illustrates some political concept, French Guiana takes the same value as a country that lies 7000 km away rather than a value influenced by neighbours.

Or consider mainland Africa. The continent has eight major physical domains, regions such as the Sahara, the savanna, the Great Lakes, and so on. These regions, according to the UN, are home to 48 countries. When we look at the continent in terms of whether a country drives on the left or right side of the road, for example, the map bears no relation to those physical regions; rather, the map exhibits a clear split derived from the events of recent human history. In some other examples that illustrate political or social points, the map of Africa takes on the appearance of a patchwork quilt.

We end this chapter, however, by looking at the world's countries through the lens of human population density. Given the above discussion, the result is perhaps surprising.

11 Countries Named After People (Map 11)

An endonym is the name people living in a particular place have given to their homeland. For example, the people living in the country that shares land borders with Norway to the north, Russia to the east, and Sweden to the west, call their land Suomi. That is the endonym for the country. An exonym is the name given to a place by people living outside that land. The English exonym for Suomi is Finland. This section examines English exonyms, which is reasonable since this book is written in English.

With English exonyms, four main classes exist: countries named after people; countries named after tribes or ethnic groups; countries named after location; and countries named after some aspect of the land. Many exonyms end in -stan, -land, -ia. These rather boring suffixes mean little more than 'country of' or 'place of'. Names in that first class of exonym, countries named after people, are more interesting.

Only a handful of large countries are named after historical figures. Even when you admit smaller nations and territories, only a further three dozen or so derive their names from people (usually men).

America and China are the two biggest countries named after historical personages. See Fig. 20. The former gets its name from the Italian explorer and mapmaker Amerigo Vespucci or, in Latinised form, Americus Vespucius. (A world map drawn in 1507 by Martin Waldseemüller appears to be the first that called the lands in the western hemisphere 'America'. The first known use of the name 'United States of America' dates back to 2 January 1776, in a letter written to the secretary of George Washington.) The latter derives from the Qin (pronounced 'Chin') dynasty, which was founded by Qin Shi Huang—the first emperor of a unified China, who was born in 259 BCE and who unified all of China in 221 BCE. He leaves a trace beyond his name: his armies were depicted in a collection of terracotta statues,

© The Author(s), under exclusive license to Springer Nature Switzerland AG 2023
S. Webb, *Around the World in 80 Ways*, https://doi.org/10.1007/978-3-031-02440-5_3

11 Countries named after people

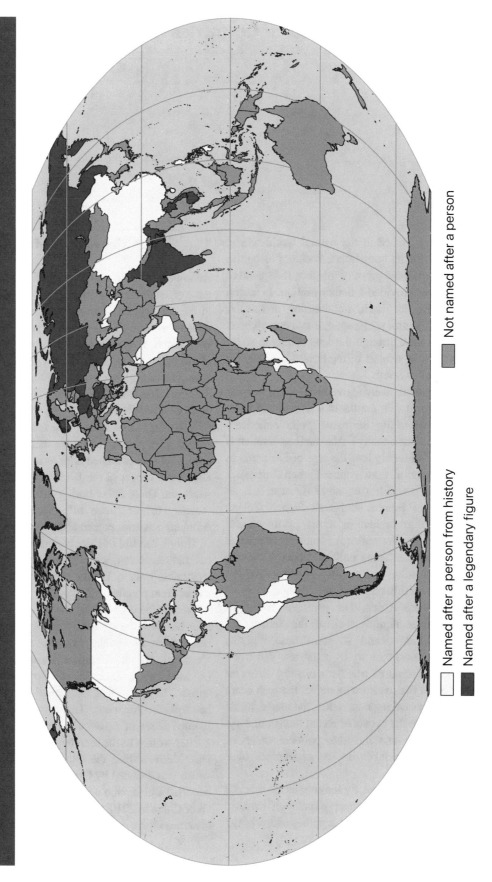

Named after a person from history

Named after a legendary figure

Not named after a person

Map 11

Fig. 20 Left: a statue of Amerigo Vespucci (Florence) (Public domain, CC0). Right: a statue of Qin Shi Huang (Xian) (JesseW900, CC BY 4.0). The two largest countries whose names derive from historical personages are America (after the explorer) and China (after the emperor)

life-size pottery figures of warriors who were supposed to guard the emperor in the afterlife. The warriors are still on display.

Slightly smaller countries with names traceable to historical figures include Colombia, Peru, and Saudi Arabia. Colombia, the 'Land of Columbus', of course honours the Italian explorer Christopher of that ilk (a European who deserves credit for reaching the Caribbean and Central and South America, but who seems to have had little idea where he actually was). Peru, at least according to one tale, gets its name from an Indian chieftain called Beru. Spanish explorers, so the story goes, asked Beru where they were; he thought they were asking for his name. This etymology is uncertain, but no more implausible than other theories about the country's name. Saudi Arabia—'Arabia of the Sauds'—refers to the founder of the country's royal family, Muhammad bin Saud, who founded the state in 1744.

Some countries have names associated with definite historical figures; some are associated with figures who, because they lived so long ago, we know little about; some are perhaps better thought of as legendary figures. The case of Mozambique, for example, is clear: it takes its name from Mussa Bin Bique, the country's ruler before the Portuguese colonised it in 1544. Jordan is named after the river, but the country's full name is the Hashemite Kingdom of Jordan. The word 'Hashemite' here relates back to Hashim ibn Abd Manaf, the great-grandfather of the prophet Muhammad. Israel, which means 'contends with God', derives its name from the patriarch Jacob, son of Isaac and Rebecca, and grandson of Abraham. Archaeologists have no evidence for his existence but, according to the Book of Genesis, Jacob wrestled with an angel and showed such perseverance that God renamed him Israel.

Nicaragua derives its name from the indigenous chieftain Nicarao who, in the sixteenth century, when the Spanish conquistadors arrived, ruled part of what is now the southwestern section of the country. Except, Nicarao seems not to have existed! Historians have recently uncovered the relevant chieftain's actual name: Macuilmiquiztli. I wonder whether, in the light of this new knowledge, the country's name will change? Less controversial is the name Eswatini, often known as Swaziland: this South African country gets its name from King Mswati II, who ruled between 1840–68.

Simón Bolívar, the military leader who helped gain independence from the Spanish Empire for Bolivia, Colombia, Ecuador, Panama, Peru, and Venezuela, lends his name to two countries: Bolivia (duh) and Venezuela. The latter more recognisably derives its name from the Spanish for 'little Venice' (at least, that is a common assumption), but the full name of the country is the Bolivarian Republic of Venezuela.

Here are a few simpler cases: the Philippines (after King Philip II of Spain); Liechtenstein (the Princely House of the same name); Seychelles (the eighteenth century French politician Jean Moreau de Séchelles); Kiribati (the eighteenth century English mariner Thomas Gilbert—'Kiribati' is how the natives pronounced 'Gilbert's'); Mauritius (Maurice, Prince of Orange—Dutch explorers found the island in 1598 and named it in his honour when he was merely Maurice of Nassau); Marshall Islands (the English explorer John Marshall); Solomon Islands (the biblical King Solomon—though, as with Israel, this stretches the definition of a historical personage); and Uzbekistan (Öz Beg Khan, whose period in charge of the Golden Horde, between 1313–1341, made him the longest-reigning khan).

In addition to Israel and Solomon Islands, religious figures from Christianity give rise to the names of several countries. For example the name El Salvador, the Saviour, refers to Jesus. Various saints make an appearance: San Marino (from Marinus, who founded a chapel there); São Tomé and Príncipe (from Thomas and the Prince of Portugal); the wonderfully named Saint Kitts and Nevis (from Christopher); Saint Lucia (from Lucy); Saint Vincent and the Grenadines (from Vincent of Saragossa); and Dominican Republic (from Dominic, whom I was surprised to learn is the patron saint of astronomers).

A number of island dependent territories take their name either from European explorers and statesmen (Bermuda, Cook, Falklands, Keeling, Norfolk, Pitcairn, South Georgia. . . I won't mention them all) or saints (Saint Helena, Sint Maarten, Saint Pierre and Miquelon—along with the Virgin Islands, named after the British Saint Ursula who is reputed to have set sail with a contingent of 11,000 virgins when she joined her future husband in France).

What about countries named after clearly legendary figures (even more legendary, that is, than Ursula and her virgins)?

The most interesting story involves three brothers who went hunting one day. They each followed different prey and ended up travelling in different directions, where they then settled. Their names gave rise to the names of several countries. The brother Rus went east. Russia is the 'land of Rus'; Belarus is 'white Rus'. The brother who went west was called Czech—and ultimately gave his name to what is now Czechia. The brother who went north was called Lech and his name gave rise to the country called Lechia, which is an ancient name still in use in some languages. Since it would be a shame to separate him from his two brothers I have coloured Lechia on the map in the same way as Russia and Czechia—but in English we call the country Poland, after the Lechitic tribe of Polans.

Romania comes from 'Roman', which ultimately derives from Romulus who, along with his twin brother Remus, is reputed to have founded the city of Rome. Armenia is a Latinised version of Hayastan, which is named after the legendary patriarch Hayk.

India is an intriguing case. The name 'India' derives from the river Indus, and has been in use for millennia. But another name in official use, and a name I have heard used increasingly in English, which is why I present it here, is Bhārat. The name presumably comes from Bhārata, which belongs to two legendary figures, either of whom might have given rise to the country's alternative name.

Two countries are named after gods: Éire (named after Ériu, a Celtic goddess of fertility) and Djibouti (which means 'land of Thoth', the Egyptian god of the Moon).

Bangladesh derives its name from 'Bengal'. According to some etymologies, Bengal derives its name from the character Bung (who, according to some theological accounts, was a great-grandson of Noah).

Norway, according to the medieval chronicle of Orkney, is said to have been founded by Nór. In an even older legend, the brothers Hunor and Magor (sons of Nimrod, another great-grandson of Noah) gave rise to the Huns and the Magyars—hence Hungary (and note that Hungarians themselves call Magyarország).

Finally, Cambodia and Laos. The former is named after Kambu Swayambhuva, a legendary king; the latter after Lava, the son of Sita, who herself was the daughter of the Earth goddess.

12   The Gender Gap (Map 12)

Since 2006, the World Economic Forum have published a Global Gender Gap Index as part of their *Global Gender Gap Report*. The 2021 Index benchmarks countries along four dimensions, comparing gender gaps in: economic participation and opportunity; educational attainment; health and survival; and political empowerment. It then combines these dimensions to give an overall score for a country's gender gap. Where women and men have parity, the gender gap is deemed to be 100% closed; anything less than 100% means a

12   The gender gap

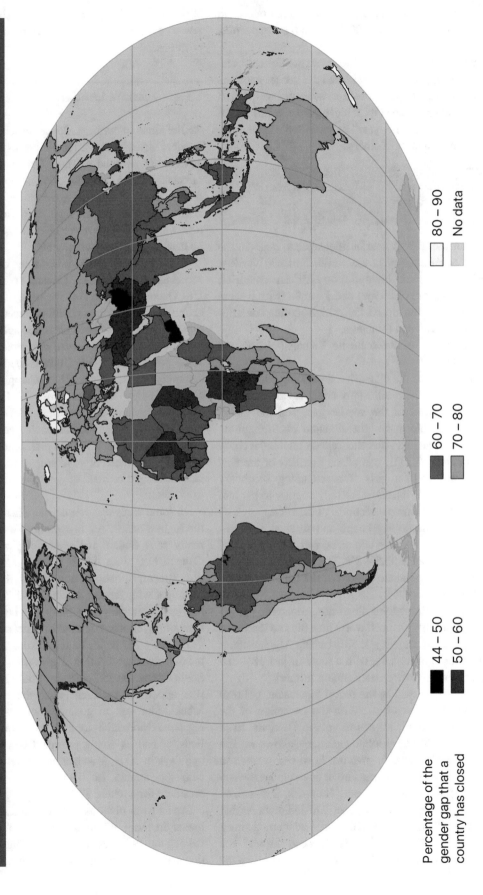

Percentage of the
gender gap that a
country has closed

44 – 50
50 – 60
60 – 70
70 – 80
80 – 90
No data

Map 12

gender gap exists. The reader can no doubt guess two main points. First, no country scores 100%. Second, the gap works in favour of men.

The effects of the Covid-19 pandemic on public health and on the economy appear to have widened the gender gap. The 2021 Index suggests that the world as a whole has covered 68% of the distance needed to achieve parity, a fall of 0.6 percentage points on the year before. Some countries, though, are more equal than others.

> On its current trajectory, it will now take 135.6 years to close the gender gap worldwide.—World Economic Forum (2021)

Iceland, for the twelfth time in fifteen years, ranks as the most gender-equal country in the world. Finland, Norway, and Sweden also rank highly. Two African countries, Namibia (with a score of 80.9%) and Rwanda (with a score of 80.5%), appear in the top-10. Other high-scoring countries include New Zealand and Lithuania.

At the other end of the scale lie Yemen (49.2%) and Afghanistan (44.4%).

The gender gap for the dimension describing educational attainment is relatively small. Many countries have entirely eliminated that aspect of the gender gap, and the global average is 95%. This makes the situation in Afghanistan even more heartbreaking. After the Taliban retook the country in the summer of 2021, one of the first acts of the new government was to ban girls from attending secondary school. Afghanistan is thus now the only country to bar half its population from gaining a secondary education.

The dimension describing health and survival also sees relatively small gaps for most countries, with the majority of scores being in the range 93.5–98%. The two most populous nations, China and India, are bottom of the list and pull the world average down to 96%.

The situation with the third dimension, economic participation and opportunity, is much worse than the two discussed above. The global average is just 58% (where Afghanistan is once again bottom of the pile, with a score of just 18% and the prospect of the situation deteriorating further).

The gender gap regarding the fourth dimension, political empowerment, is worst of all. With the exception of the Vatican, all states permit women to vote in elections. Even Saudi Arabia, which is an absolute monarchy, allows women to vote in local elections. In practice, however, women in many countries face societal pressure or even violence when they try to cast a vote. And in many countries, few women ever appear as a name on a ballot paper. The Index shows that Vanuatu and Papua New Guinea had no women parliamentarians; of the parliamentary seats around the world, women occupied only 26.1% of them rather than the 50% one would expect. And, as we shall see with Map 13, few women in the world rise to occupy the highest political positions.

13 Female Leaders (Map 13)

At the time of writing, 25 countries—about 13% of UN member states—have a female elected or appointed head of government or head of state. One would hope that, at any moment in time, roughly 50% of countries would have a female head of state or government. This, of course, has never been true. As the map shows, many countries in the post-war period have *never* elected or appointed a woman to be head of state or government.

A map such as this requires some clarification. Consider, for example, Australia, Canada, New Zealand, and the UK. Queen Elizabeth II is the head of state of each of these countries, but hers is a position gained by birth. The reason these countries have been classified as having a female leader is that they have each *elected or appointed* a female leader at some point since 1950. (Julia Gillard was the first female Australian prime minister; Kim Campbell the first Canadian; Jenny Shipley the first New Zealander; and Margaret Thatcher became the first female UK prime minister.) In other words, a female monarch does not count for our purposes here; but a female president (who might be head of state and/or head of government) or a female prime minister (who might be head of government but not head of state) does count.

In some cases one can debate whether a woman has held the highest office. An 11-member Sovereignty Council currently runs Sudan, for example. Two women sit on the Council but, since this is a short-term, collective form of government, I choose not to put Sudan in the rank of countries with female leaders. And Namibia, for example, has had a female prime minister—but the office is neither head of government or head of state, so I exclude it too.

On the other hand, I am generous to several countries in making my classification. I give the benefit of the doubt to South Africa. On 25 September 2008, Ivy Matsepe-Casaburri was appointed president after the resignation of Thabo Mbeki. She was in post for only 14 h, however, before Kgalema Motlanthe took office. For two days in 1997, Ecuador had an acting president who was female and, for two days in 2009, Madagascar had an acting prime minister who was female. In 2018, Vietnam had a female acting president for a whole 32 days.

Apart from obvious issues regarding equality, does this matter in terms of good governance? In other words, are women any better than men at leading a country? The question is perhaps meaningless, but writing from a vantage point

13   Female leaders

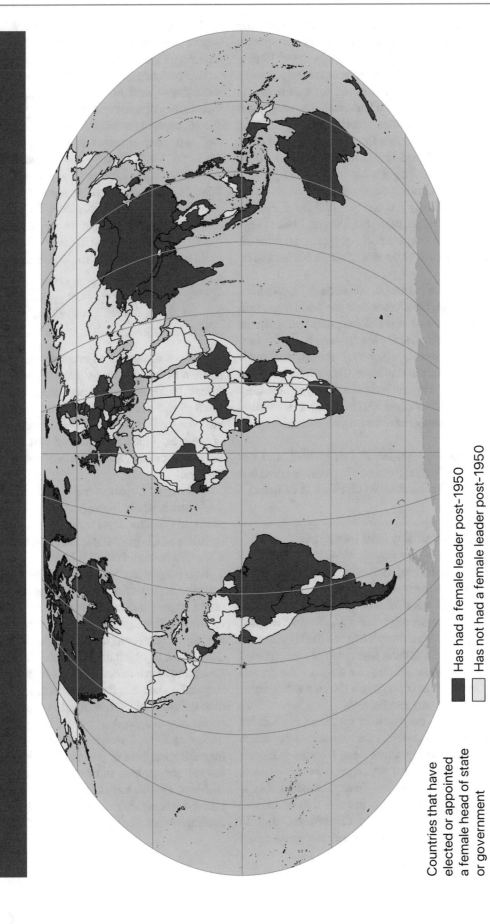

Countries that have
elected or appointed
a female head of state
or government

■ Has had a female leader post–1950

□ Has not had a female leader post–1950

Map 13

midway through a pandemic I can't help wondering whether countries led by women enjoyed a more competent response.

The Worldometer website collates various Covid-19 data sets from all countries, and thus permits country-by-country comparisons. One straightforward metric for comparison is Covid-19 deaths per million of population. The comparison is not perfect because countries report deaths in different ways. (Some states report deaths *from* Covid-19 while some report all deaths *with* Covid-19.) Furthermore, some places possess geographical advantages when it comes to fighting a pandemic. (Tristan da Cunha is the remotest settlement on the planet, and unsurprisingly has not yet suffered a case of Covid-19; Germany, on the other hand, shares borders with nine countries.) But deaths per million of population provides at least a rough indication of the size of the problem faced by a country.

At the outset of the pandemic, women led 17 countries: 11 of them have so far fared better than the world average in terms of deaths per million of population; only two are in the top-60 worst-hit countries. Compare that with countries such as USA, UK, and Brazil, which had stereotypical male leaders at the onset of the pandemic. This correlation does *not* demonstrate causality. For example, the type of society that elects female leaders might also be likely to have dealt well with Covid for a range of other reasons. But it does offer pause for thought.

(Note added: Queen Elizabeth II died in 2022, and was succeeded by King Charles III. Rules of succession mean that at least the next two monarchs after Charles will be male.)

14 Flags Containing Red (Map 14)

Considering the billions of dollars the corporate world lavishes each year on shiny new logos and bespoke colour schemes, it surprises me how little money national states devote to their visual identity. The CIA World Factbook displays an image of each country's flag and, flicking through them, I have to say most of them look much of a muchness. I have real difficulty distinguishing the flags of Russia and Slovenia, for example, or (much to my German wife's irritation) the flags of Germany and Belgium.

My difficulty no doubt stems in part from my lack of visual awareness. But a glance at the map shows the dearth of imagination employed by states when it comes to colour: most flags contain red. Indeed, four countries—China, Morocco, Turkey, and Vietnam—have a standard that is more than 90% red. China's flag has five yellow stars to break up the field of red; the ruler of Morocco, back in 1915, chose a green pentagram on a red background (not

ideal for those suffering from Daltonism); a white star and crescent leavens the red of Turkey's flag; and Vietnam goes for a single yellow star on the red background.

Red probably arises for a variety of reasons. In Europe, some flags—such as those for England and Spain—have roots in heraldry; red was the most common heraldic tincture. The red cross of the English flag made its way into the Union flag, and many countries went on to derive their own flag from the UK's. Many Slavic countries allude to the pan-Slavic white–blue–red, which in turn is based on Russia's flag (where red represents the people). And of course red often represents blood: the Chinese and Vietnamese flags ultimately symbolise the blood spilled during the French revolution, the red in many other flags symbolises the blood spilled in struggles for independence.

> There is hopeful symbolism in the fact that flags do not wave in a vacuum.—Arthur C Clarke

In terms of number of occurrences, white appears almost as often as red on national flags. In terms of area coverage, though, blue is the second most popular colour. If you were to stitch all national flags together to form a 'flag of the world' then you would see blue and white each covering about 20% of the area, with the amount of blue being slightly greater than that of white. Some shade of red would cover 30% of the area.

The flag of Cyprus is the whitest: more than 92% of the flag's area is white. Japan, South Korea, and Israel also have noticeably white flags. And four countries—Democratic Republic of Congo, Micronesia, Nauru, and Somalia—have a flag that is more than 90% blue.

Some states are not formally recognised. These states nevertheless have flags, and red appears as often on these as on the flags of universally recognised nation states. The flag of Western Sahara, for example, contains a red triangle and a red star-and-crescent; the flag of Somaliland has three horizontal bars, of which the bottom bar is red; the flag of South Ossetia has three horizontal bars, of which the middle bar is red; and so on. The flag of Antartica is more interesting: it does not exist. At least, no flag has general recognition. If the continent ever agrees on an official flag, then the 'True South' design of Evan Townsend, created in 2018, might be it: increasing numbers of programs use it. The True South flag has two stripes, blue and white, representing the long nights and days; a white mountain rising into the blue stripe casts a blue shadow in the white stripe. It is a beautiful emblem. And contains no red.

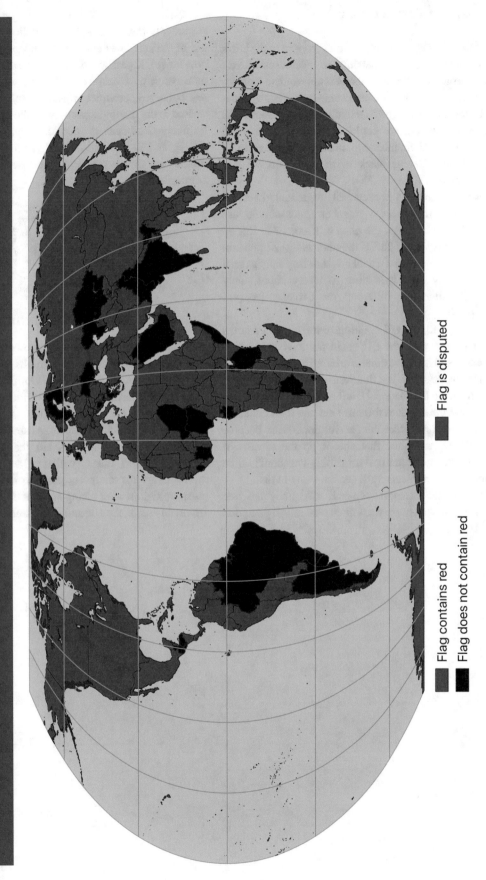

14   Flags containing red

Flag is disputed

Flag contains red

Flag does not contain red

Map 14

15   Passport Colours (Map 15)

Of all the specious arguments raised in the run-up to the UK's 2016 referendum on EU membership the speciousest—and I believe the barbarity of this neologism is appropriate, considering the quality of the arguments made—was that, by leaving the EU, the UK would be free to choose the colour of its passport. We were told we could ditch the burgundy-red passport that our EU membership mandated and return to the midnight blue (seems black to me) passport we had proudly used since 1921.

This was always a nonsensical argument because, although the EU *recommends* member states should use burgundy-red for passports, the choice of colour is left up to individual countries. Of 28 EU member states, 27 presumably decided this was such a trivial concern they may as well go with burgundy; one less thing to worry about. Only Croatia chose something different: the Croatian passport remains dark blue.

So, had it wished, the UK could have stayed with its own dark blue passport. And the EU could not have cared less. What the UK could *not* have done is arbitrarily change the size and format of its passport. But that restriction comes not from the EU but from the International Civil Aviation Organization. The idea behind the standard format is simple: if all countries adhere to a common format then travel documents become machine readable, a feature that speeds up the process of entering and leaving a country—something we should all be grateful for. Even Brexiteers must see the value in that.

Considering that passport colour is essentially the free choice of a nation, we see only a limited palette. All passports are shades of just four colours: red, green, blue, and black. Admittedly there can be large variation in shading. The reds are of various hues and, as mentioned above, the old (and now new) UK passport is of a blue so deep I and many others perceive it as black. Nevertheless, we do not see yellow passports, say, or violet or grey passports.

Red is the commonest colour, followed by blue (favoured by many Caribbean countries), green (common in Islamic states), and black. The lack of variety might ultimately be rooted in a simple practicality: only a few companies have permission to produce passports.

Some passport facts:
Flick through a Finnish or Slovenian passport and images along the bottom of the page create a moving picture.
The US Department of State says that most US citizens don't have a passport.
The Nicaraguan passport, with 89 security features, is the most difficult to forge.—BBC (2017)

Passport Index, a website that shows the colour and design of all the world's passports, also ranks passports according to the total mobility they offer. On this scale, Japan and New Zealand have the most powerful passports; in the second rank are the passports of a number of European countries, along with those of Australia and South Korea. Afghan and Iraqi passports are the least powerful. The Henley Passport Index, which offers a slightly different methodology for analysing passport power, rates the Japan passport highest after taking account of the impact of Covid-19; on this scale, the UK and US passports are in the 7th rank.

15 Passport colours

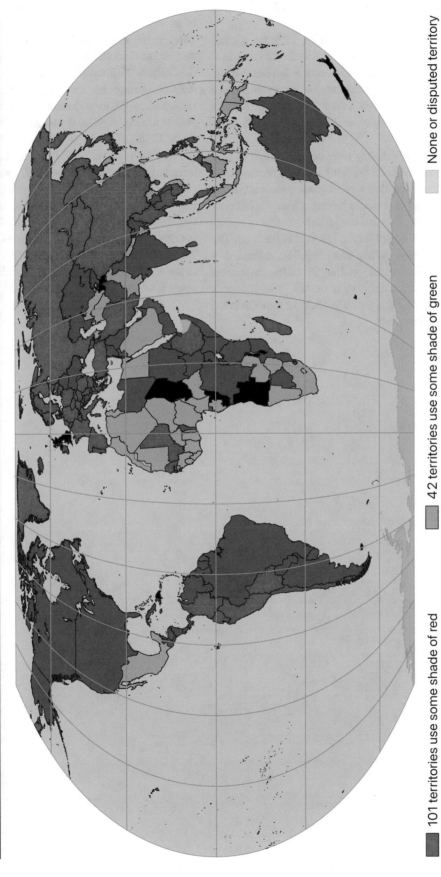

101 territories use some shade of red

89 territories use some shade of blue

42 territories use some shade of green

14 territories use black/blue–black

None or disputed territory

Map 15

16   Driving: Left or Right? (Map 16)

My German wife has lived in England for 20 years yet still complains about driving on the left. She makes it seem as if the UK is the only place where traffic flows in left-hand lanes, but countries that drive on the left account for about 25% of the world's roads and contain about 35% of the world's population. She raises a question, though: why *do* Brits drive on the left?

A plausible story, which may even be true, is that sword-carrying horse riders *rode* on the left-hand side because they naturally wore the scabbard on the left (most people are right-handed). Mounting a horse would be easier from the left and, since this is safer at the side of a road than in middle of traffic, it makes sense to ride horses on the left side of the road. The habit became ingrained. When road building expanded it became necessary to regulate traffic; in 1835, Parliament made driving on the left mandatory in the UK (with countries in the Empire following this lead). When cars were invented (see Fig. 21) they drove left.

Contrast this with France and the USA. Farmers used pairs of horses to pull wagons; the driver sat rear-left so the right arm was free to lash the horses. In this case oncoming wagons should pass on the left so both drivers can ensure wheels don't clash. Wagons thus drove on the right. Napoleon spread the rule through his conquests, and the newly independent USA had no desire to follow British customs (Pennsylvania passed a 'keep right' law as early as 1792).

In 1918, as many territories drove left as drove right. Seventy years later, 34 had switched from left to right (only three have gone the other way). The main reason was desired or forced harmonisation with neighbours. Portugal, for example, switched in 1928; Spain followed in 1930. Austria and Czechoslovakia both switched in 1938. Sweden, the last country in continental Europe to switch, did so on 3 September 1967, the so-called 'day N'. My wife is unlikely to see a 'day N' switch in the UK: with such a high density of roads it would cost too much.

16   Driving – left or right?

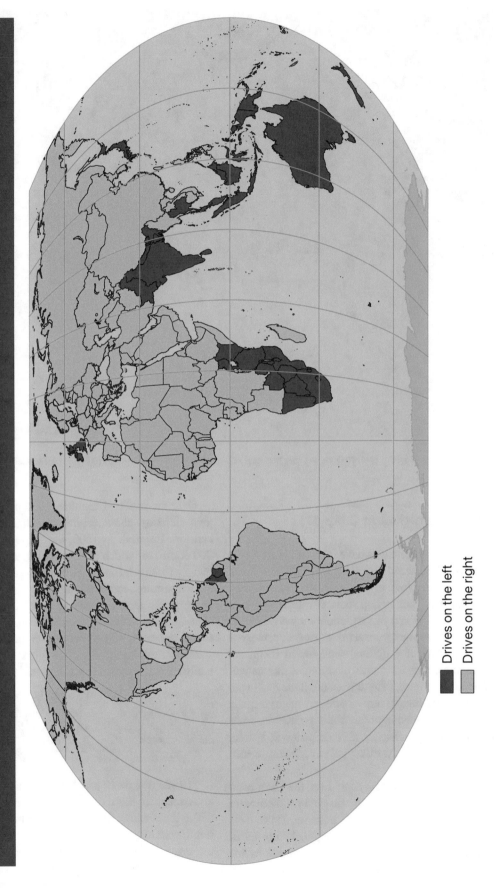

Drives on the left
Drives on the right

Map 16

Fig. 21 A replica of Benz's Patent Motorwagen, the first practical self-propelled vehicle, patented in 1886. (Alexander Migl, CC BY 4.0)

17   Capital Punishment (Map 17)

For most of history, communities have found it acceptable to execute those who threatened societal norms and those found guilty of serious crimes (see, for example, the discussion on homicide in Map 51). Until nation states developed a proper prison system—and the sort of institution we would recognise as a prison did not emerge until the nineteenth century—judges had few sentencing options available to them, not many ways of preventing recidivism, and limited alternatives when it came to deterrence. Petty crimes might be dealt with through corporal punishment; serious offences might result in banishment or transportation to colonies (which in some cases might be the equivalent of a death penalty); but for rape, murder, and treason. . . well, capital punishment seemed the natural solution.

Capital punishment derives its name from the Latin *capit* meaning 'head'. The term refers to one obvious method of execution: beheading. Over the years, however, states have adopted numerous methods of execution, including lethal injection, hanging, electrocution, and shooting. Saudi Arabia is now the only country that still beheads people as an official punishment.

Over the past two centuries most societies have changed their thinking about capital punishment. As an example, consider England. In 1800, the 'Bloody Code' listed about 200 crimes for which death was the statutory punishment. A hungry teenager, found guilty of shoplifting a crust or stealing a few coins, could be hanged. If such a case came to trial, many jurors would refuse to find the suspect guilty. And some judges, when it came to cases of petty theft, valued stolen goods at a level below the amount that triggered a statutory death penalty. In a situation such as this, law eventually catches up with evolving societal values.

> The death penalty—in violation of international law:
> At least one *public execution* was recorded in Iran.—Amnesty International (2021)

In 1823, the death penalty in England became discretionary, except for those found guilty of treason and murder, for which the statutory punishment remained death. As the nineteenth century progressed, the number of offences for which the death penalty was a possible sentence diminished. In 1832, for example, thieves, counterfeiters, and forgers knew they could not be hanged for their misdeeds (unless they

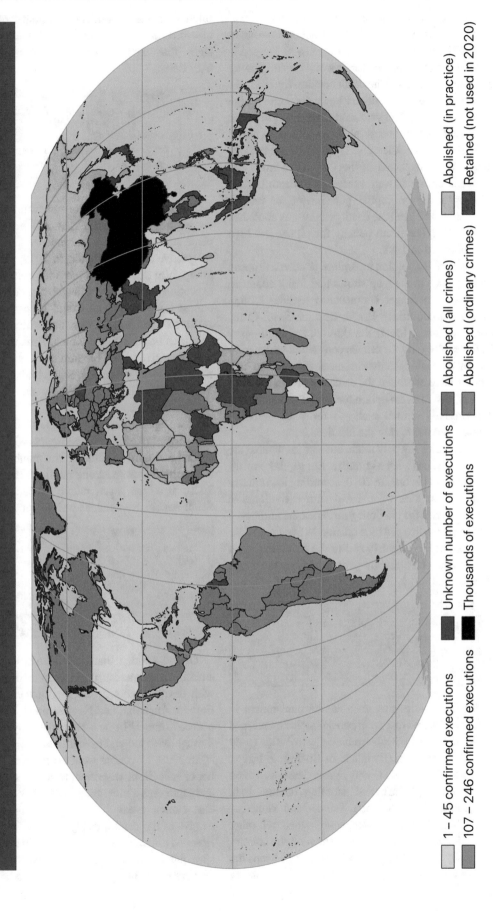

17   Capital punishment

1 – 45 confirmed executions

107 – 246 confirmed executions

Unknown number of executions

Thousands of executions

Abolished (all crimes)

Abolished (ordinary crimes)

Abolished (in practice)

Retained (not used in 2020)

Map 17

forged wills or powers of attorney; the death penalty was retained for those crimes). In 1835, two Londoners—James Pratt and John Smith—became the last men to be hanged in England for sodomy. In 1837, the forgery of wills and powers of attorney no longer attracted the death penalty. By 1861, the only capital crimes outside of military law were murder, high treason, arson in a Royal Dockyard, and piracy. As the prison service developed, public unease about capital punishment increased. In 1964, Peter Allen and Gwynne Evans became the last people to be executed in the UK. After those two hangings, several criminals were sentenced to death but not hanged: a moratorium on capital punishment for murder came into effect in 1965, and the death penalty was abolished for murder in 1969. The death penalty for piracy and high treason was abolished in 1998.

A similar pattern of abolition took place in other European nations, and a glance at the map shows that many countries around the world no longer countenance judicial killing. Furthermore, eight countries—Brazil, Burkina Faso, Chile, El Salvador, Guatemala, Israel, Kazakhstan, and Peru—permit the death penalty only for crimes committed under military law or in exceptional circumstances. More than two dozen countries have abolished the death penalty in practice: in these nations, although the option for capital punishment is retained for 'ordinary' crimes such as murder, the sentence has not been carried out during the last decade.

Dozens of countries not only retain but also implement the death penalty for crimes such as murder. But capital punishment appears to be in decline. In 2020, according to Amnesty International, 18 countries went ahead with executions—a drop from 20 countries the previous year. The total number of confirmed executions dropped by a quarter. While the Covid-19 pandemic might have contributed to the reduction, each year since 2015 has seen a year-on-year decrease in the use of capital punishment.

> The death penalty—in violation of international law:
> Three people in Iran were executed for crimes that occurred when they were *below the age of 18*.—Amnesty International (2021)

Among those countries that used capital punishment in 2020, five methods were employed: beheading (as mentioned earlier, this is used only in Saudi Arabia); electrocution (used only in the USA); hanging (Bangladesh, Botswana, Egypt, India, Iran, Iraq, South Sudan, and Syria); lethal injection (China, USA, and Vietnam); and shooting (China, Iran, North Korea, Oman, Qatar, Somalia, Taiwan, and Yemen).

In 2020, China executed more people than all other nations on Earth combined, a dubious distinction the country has held for many years. The Chinese government keeps the number of executions a state secret, but Amnesty

International are confident the number of judicial killings is measured in the thousands. Even in China, however, there are signs that the use of capital punishment is declining over time.

> The death penalty—in violation of international law:
> People with *mental or intellectual disabilities* were known to be under sentence of death in several countries, including Japan, Maldives, Pakistan, and USA.—Amnesty International (2021)

The number of executions in North Korea is a matter for conjecture. Amnesty International believe executions take place after unfair trials, but independent verification in such a secretive state is difficult. In 2019, for example, a justice group from South Korea provided evidence of 318 places of execution in its northern neighbour, but confirmation proved almost impossible. The Covid-19 pandemic meant that, in 2020, the number of executions was imprecise in several countries besides North Korea. Amnesty International were unable to give a credible minimum number of executions in Iran, Egypt, Iraq, Somalia, South Sudan, Syria, Vietnam, and Yemen. The numbers given by the organisation are thus minimum estimates.

If we ignore the secretive North Korea, then Iran (with 246 verified executions) and Egypt (with 107 executions) are second and third to China in the number of judicial killings. Iran permitted at least one public execution to take place in 2020; in recent years, public executions have occurred in Saudi Arabia, Somalia, as well as (of course) North Korea.

> The death penalty—in violation of international law:
> Death sentences were known to have been imposed after proceedings that did not meet *international fair trial standards* in several countries, including Bahrain, Bangladesh, Egypt, Iran, Iraq, Malaysia, Pakistan, Saudi Arabia, Singapore, Vietnam, and Yemen.—Amnesty International (2021)

Of the nations that retain the death penalty, the USA attracts most interest: the country appears to be out of step with developments in the rest of the Americas as well as Europe, Australia, and New Zealand. Indeed, for more than a decade the USA has been the only such country to go through with judicial executions.

In 2020, 17 people were executed in the USA. The number of executions was down from 22 in 2019 and 25 in 2018, and represents a more than fourfold decrease in just 20 years. One cannot predict, based purely on these numbers, the demise of the death penalty in the USA: federal executions *resumed* after a 17-year hiatus, and in a six-month period the state put ten men to death (compared to just three federal executions in the preceding four decades). The low overall figure for executions was a consequence of the Covid-19 pandemic: many states that routinely execute prisoners were badly affected by the coronavirus and many court proceedings had to be put on hold. Nevertheless, over a longer timescale, one can perhaps discern a trend towards abolition. By the start of 2021, 22 out of the 50 US states had abolished capital punishment for all crimes; a further 12 states have not implemented the punishment for more than a decade.

> The death penalty—in violation of international law: "Confessions" that may have been extracted through *torture or other ill-treatment* were used to convict and sentence people to death in Bahrain, Egypt, Iran, and Saudi Arabia.—Amnesty International (2021)

In part, this reduction in capital punishment in the USA might be due to a public recognition of the injustices that are still taking place. Two recent cases in particular aroused concern.

In February 2019, the state of Alabama put Domineque Ray to death for the murder twenty years earlier of a teenage girl. At his trial, however, the jury were unable to reach a unanimous verdict because the prosecution provided no physical evidence that linked him to the murder. And in August 2019, the state of Texas put Larry Swearingen to death for the murder of a teenager in 1998; prosecutors again relied on circumstantial evidence. In North Carolina, on the other hand, two men who between them had served 84 years in prison, were released from death-row when the system realised they were innocent.

Since 1985, the Gallup organisation has asked the American public about the death penalty. In 2019, for the first time, most Americans answered that they favour the sentence of life imprisonment over the death penalty.

18   Press Freedom (Map 18)

Article 19 of the UN's *Universal Declaration of Human Rights* states that "freedom of opinion and expression" implies the right to "seek, receive and impart information and ideas through any media and regardless of frontiers". The Article expresses a fine sentiment but, as we all know, many people around the world struggle to exercise that basic right. (At least, those of us who have ready access to information know the difficulties faced by others. Censorship in North Korea is so strict that the country's citizens will perhaps struggle to grasp the extent to which their right to seek information has been curtailed compared to others.)

Reporters Without Borders, a Paris-based group that derives its acronym RSF from its French name, Reporters Sans Frontières, is an independent organisation that aims to protect press freedom and independent journalism. One way it does this is by compiling and publishing the annual *World Press Freedom Index*. Through its offices in ten cities across the world, and its network of correspondents in 130 countries, RSF asks experts to look at points such as the independence of a country's news media; the level of plurality of news sources; the quality of the legislative framework in this area; and the safety of journalists. (Note: the quality of journalism itself is *not* ranked.) The group combine this qualitative data with quantitative data on abuses and acts of violence against journalists. RSF as an institution possesses significant heft, holding consultative status at the United Nations, UNESCO, Council of Europe and International Organization of the Francophonie; its *Index*, which has been published since 2002, has a growing influence across the world, and organisations such as the UN and the World Bank refer to it.

So what does the *Index* say for 2021, the most recent data to which I have access?

The *Index* ranks countries into five groups, ranging from 'Good' (a category into which 12 countries fall) to 'Very bad' (21 countries).

Countries in the 'Very bad' category constitute the usual suspects. China has a poor record on freedom of expression, of course, and journalists have a tough time plying their trade in places such as Syria, Libya, and Iran. For several years, the Central Asian country of Turkmenistan has floundered in the lower reaches of this league table: its government controls all media, provides access to only a highly-censored version of the internet, and has had journalists based abroad harassed and even tortured. It will be of no surprise to anyone that North Korea lies near the bottom of this ranking. What is more surprising is that, in 2021, RSF finds reason to rank a country below North Korea: Eritrea. For two decades, journalism in Eritrea has been suppressed; at least 11 reporters are in prison for doing their job, and have no access to family or legal representation; Eritreans even struggle to access credible news via radio (the single independent radio station often has its signal jammed) or internet (citizens must provide ID before accessing an Internet cafe).

At the opposite end of the scale, northern countries such as Norway, Finland, Sweden, and Denmark all possess an admirable record. But even in an open country such as the Netherlands, which ranks sixth on the *Index*, journalists can encounter problems. In 2020, a news year dominated by Covid-19, Dutch journalists were routinely attacked when reporting on illegal gatherings and lockdown protests. Furthermore, intimidation from organised crime remains an ongoing threat: one investigative reporter found a hand grenade outside his house, and two journalists working on stories related to organised crime are now living under permanent police protection.

Freedom of the press is a right worth protecting.

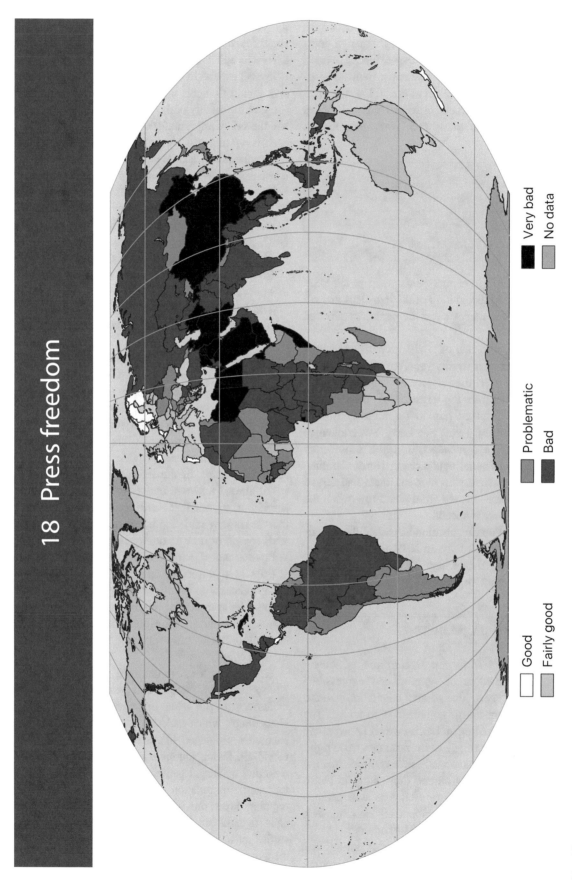

18 Press freedom

Good

Fairly good

Problematic

Bad

Very bad

No data

Map 18

19   Dominance of a Capital City (Map 19)

For most of our species' time on this planet, people have generally lived in environments with a low population density: humans roamed the land as hunter-gatherers. The invention of agriculture permitted people to live in fixed locations, and around 7500 BCE the first cities—Eridu, Ur, Uruk, and similar settlements—were founded in Mesopotamia. But even nine millennia later the majority of humans lived in non-urban environments: in 1900, just 16% of people lived in towns and cities. Only later in the twentieth century did urbanisation boom. In 2007, for the first time, more humans lived in urban than in rural areas.

Will we ever reach the stage where Earth resembles Isaac Asimov's fictional Trantor—the domed, skyscrapered, overpopulated galactic capital that depends upon other planets for alimentation? When I visit London I sometimes get the feeling we will. But London is clearly different to the rest of the UK: the city seems to suck up the country's resource. And that got me wondering: what fraction of the UK population actually lives in London? And is that fraction much greater than for the capital city of other countries?

When trying to answer this question I hit a couple of problems.

First, a handful of countries have more than one capital city. South Africa, for example, has separate administrative, judicial, and legislative capitals (Pretoria, Bloemfontein, and Cape Town respectively). In these cases I have taken the capital with the most people. Second, and more problematic, how should the scope of a city be defined? For example, when we talk of 'London' in this context should we restrict ourselves to the City of London (which is undoubtedly important, but tiny)? Central London (which perhaps is what tourists want to visit)? Or Greater London (which, confusingly, contains within it two cities—the City of London and the City of Westminster—but is not itself a city)? For this second problem I turned to the UN's *World Urbanization Prospects* for guidance.

The UN document discusses the notion of a 'metropolitan area', which is based on the idea of the labour market area of a city. That is precisely what I had in mind when thinking about the influence of 'London' on the UK, and so I used the UN definition. Be aware that there is no consistent definition across countries of what constitutes a metropolitan area. For what it is worth, though, according to the UN the world's largest metropolitan area is Tokyo, which is a cluster of 37 million souls. Tokyo is followed by Delhi (29 million) and Shanghai (26 million). London comes 37th on the list, with a population of just over 9 million.

If you order capital cities in terms of their population relative to their country's population then, at the top of the list, you get the result you might expect: some places (Gibraltar, Monaco, and the Vatican, for example, shown here by dots) are essentially city states. Large countries—China, India, and Russia, for example—have huge cities and yet the bulk of the population still live outwith the capital. Europe is more interesting.

It turns out Paris is home to a greater proportion of French citizens than London is to Brits. Just over 13% of the UK population lives in London; just under 17% of the French population live in the Paris metropolitan area. Even if we restrict the discussion to England rather than the UK as a whole, London is still not as dominant a presence as Paris. (So my intuition about London, at least in terms of population, was wrong.) Neither London nor Paris are as dominant in their respective countries as Lisbon is to Portugal. But the picture is different in some other major European capitals. Only 7% of the Italian population live in the Rome area, for example; and only 4% of the German population lives in the Berlin area.

Across the world, the least dominant capital on this measure is Dodoma. Far fewer than one in 200 Tanzanian citizens live in their country's official capital city.

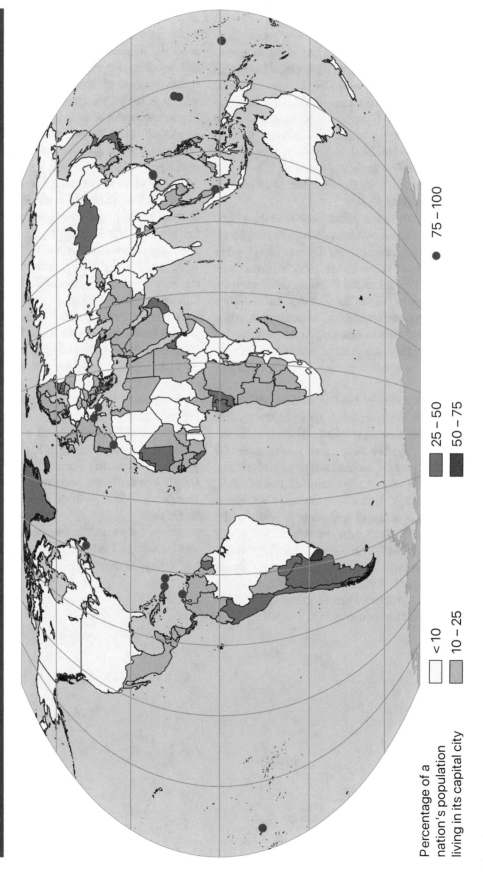

19 Dominance of a capital city

Percentage of a
nation's population
living in its capital city

☐ <10   ☐ 25 – 50   ● 75 – 100

☐ 10 – 25   ■ 50 – 75

Map 19

20   Population Density (Map 20)

A meaningful country-by-country comparison of human population density—the average number of humans per square kilometre—is difficult to present using a world choropleth. The problem is the vast range that must be represented. The population density of many countries is surprisingly low. A handful of countries and regions, on the other hand, have a high population density. Depicting all that diversity on a choropleth map is tricky.

On this map I use red dots to represent the five territories with the highest population densities. (You might only be able to make out four red dots. That is because two of the territories are geographically close.) These five hyper-concentrated pockets of humanity are:

Macau—a special administrative region of China. The 'Las Vegas of Asia' (see Map 77) is home to about 680,000 people, who enjoy one of the highest per capita incomes in the world (see Map 75) and a high life expectancy (see Map 60). The inhabitants live on just 32.9 square kilometres, most of which is land reclaimed from the sea. With 20,479 people per square kilometre, tiny Macau is the most densely populated region on Earth. See Fig. 22.

Monaco—the second smallest country, after the Vatican, is home to 19,348 people per square kilometre. The Principality of Monaco is thus the most densely populated sovereign country on Earth. The country, of course, is synonymous with wealth and expensive living.

Singapore—the sovereign island city-state has a population density of 7916 people per square kilometre. As with Macau and Monaco, the citizens of Singapore enjoy a high per capita income. Singapore is significantly less crowded than Macau and Monaco, but is nevertheless third on the list.

Hong Kong—this is another special administrative region of China. I always think of Hong Kong as being an extremely crowded place, but with a population density of 7039 people per square kilometre it is much less densely packed than Macau.

Gibraltar—the 32,000 inhabitants of this British Overseas Territory occupy a town at the foot of the Rock, which gives rise to a population density of 3457 people per square kilometre.

Even when we omit these five outliers we still face a problem when generating a world choropleth of population density: Earth is home to about 7.8 billion people but, as of 2020, our planet as a whole remains sparsely populated. Although a few countries support large population densities, most have a population density of less than 200 people per square kilometre and some are sparsely populated indeed.

Greenland is the most sparsely populated country on Earth. (As usual, we can exclude Antarctica from the discussion. Antarctica has never had an indigenous human population. Although people work at various research stations, as shown in Map 1, the total population is never more than a few thousand.) Greenland is the world's twelfth largest country in terms of area but it has only 56,673 inhabitants. Its population density is just 0.1 person per square kilometre, with most Greenlanders choosing to enjoy the relatively milder climate of the southwestern fjords and avoiding the inhospitable ice-covered areas. See Fig. 23.

Mongolia, which covers a slightly smaller area than Greenland, is home to just over three million people. About half of Mongolian citizens reside in Ulaanbaatar, the capital city (see Map 19). Mongolia is the most sparsely populated sovereign state in the world, with fewer than 2 people per square kilometre living there. Most of the country is uninhabited: mountains lie to the north and west, the Gobi desert lies to the south, and grassy steppe covers much of its area. Although people have lived in Mongolia for 40,000 years or more, humans have yet to tame the country.

Some of the largest countries have tiny population densities. The figure for Australia is 3.2 people per square kilometre; for Canada it is 4.04 people per square kilometre; for Russia it is 8.8 people per square kilometre. These numbers are thousands of times smaller than those for Macau and Monaco. Of course, Macau and Monaco are tiny places but the population densities of those three large countries—Australia, Canada, Russia—are small even compared to the average population density for Earth. The total area of Earth is just over 510 million square kilometres, of which the land area is just under 150 million square kilometres. Assuming a total population of around 7.8 billion souls, the average population density on land is about 52 people per square kilometre. The country with the closest population density to this global average is Guinea. But the USA has a smaller figure than this (35.6 people per square kilometre); so does Brazil (25.04 people per square kilometre); and so does Argentina (16.18 people per square kilometre). That is why most of the map has been assigned a pale colour.

So where does the common notion arise that the world is 'overcrowded'?

Well, a glance at the map demonstrates that certain countries in western Europe, for example, *are* densely populated compared to the large countries mentioned above. The population density of the Netherlands is 508.5 people per square kilometre, which is more than 14 times greater than the figure for the USA; the population density of Great Britain is 272.9 people per square kilometre, which is almost 11 times greater than the figure for Brazil; Germany has a population density of 236.7 people per square kilometre, which is almost 15 times greater than the figure for Argentina. So, compared to the sparsely populated countries with large land areas, the countries of western Europe might feel (and in reality are) crowded.

20 Population density

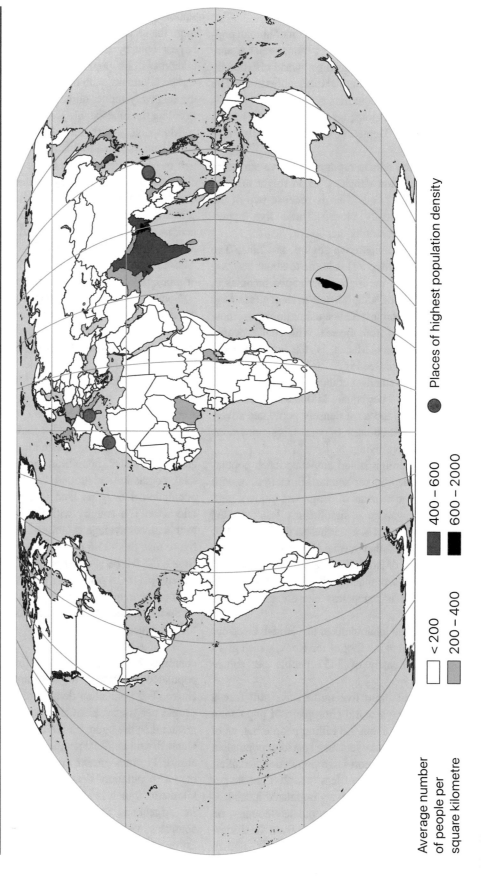

Average number
of people per
square kilometre

Map 20

| | < 200 | | 400 – 600 |
| | 200 – 400 | | 600 – 2000 |

● Places of highest population density

Fig. 22 A view of Macau looking out from the Macau Museum. The image gives an impression of the crowded nature of life in the region: residents have built high wherever possible. (AndyHe829, CC BY 3.0)

Certain other countries—Bangladesh, Korea, Taiwan, and numerous island states such as Bahrain, the Maldives, and Malta—are even more densely populated than the Netherlands.

But compared to the most densely populated regions on Earth, all the countries mentioned above still provide lots of space for their inhabitants. Indeed, that leads to an interesting question: if all of Earth's 7.8 billion inhabitants were squeezed onto a single island such that the population density of that island were the same as Macau—20,479 people per square kilometre—how big would the island be? The answer, perhaps surprisingly, is that the island would have about 65% of the area of Madagascar.

The map shows an imaginary island to the southeast of Madagascar, coloured in black, upon which all humanity could in principle fit—if we had the same amount of personal space as the people of Macau.

So—is overpopulation something we need not worry about? The answer, of course, is that we really *do* have to be concerned about the carrying capacity of our planet.

Earth as a place for human habitation is under stress. The more humans there are, the greater is the area of land required

Fig. 23 A view over Ilulissat, the third largest city in Greenland. As of 2020, the town's population was 4670. Most of Greenland is uninhabited. The difference with Macau is stark. (Lucas Cullen, CC BY 3.0)

to produce food; the greater is the demand on water for drinking, bathing, and manufacturing; the greater is the requirement for energy, a requirement that for so long has been met by burning fossil fuels (Map 21). And as our activities cause the planet to warm, the more difficult it will be to provide food and water and energy for increasing numbers of people. Science has been almost miraculously successful in providing sufficient food for the billions of mouths on the planet. (That not all mouths receive the food is not the fault of science.) But in a warming world will we be able to feed 9 billion mouths? 10 billion? More?

This is a difficult question to answer because different countries have differing standards of infrastructure, diverse resources, varying levels of agriculture; some countries will handle a growing population more successfully than others. But people are increasingly mobile. The world population is predicted to reach 9.7 billion by around 2050, before falling.

By 2050 it seems likely that global warming will have started to make the equatorial regions increasingly inhospitable, and even regions far from the equator can expect to experience episodes of water scarcity. In that scenario, what will happen to the colours on the map here? My guess is there will be a mass migration of people from south to north. Some would argue that we saw a glimpse of this in 2015. That year, more than a million migrants crossed into Europe. The main driver of migration was the conflict in Syria. Some researchers argue that the extreme drought in Syria between 2006 to 2009, a drought likely exacerbated by anthropogenic climate change, was a major cause for the uprising that began in 2011. This influx of refugees caused division in Europe over how best to respond. The number of climate refugees in coming decades is likely to dwarf the numbers seen in 2015.

The World Perturbed by People

Or planet can seem overwhelmingly large, so large as to make our actions appear inconsequential. But Earth is now home to 7.8 billion people, and humans have the capacity to shape the world in profound ways. If our wisdom matched our intelligence then we might be confident of our future. Unfortunately, improvements in our judgement have not kept pace with improvements in our technology. Instead, as the natural historian David Attenborough has noted, humanity might best be described as a "plague on the Earth".

With 7.8 billion people needing food and water, and demanding power for heating and lighting and transport, the question is not *whether* humanity is disturbing the natural world but by *how much* we are disturbing it. The maps in this chapter illustrate a few of the many impacts we have. Identifying the adverse effects of our technological civilisation is easy; finding solutions is hard.

21 Carbon Dioxide Emissions (Map 21)

Climate change is the gravest problem facing our species. Human activity is warming the world and the consequences could be catastrophic. The *world* will continue: Earth has been hotter in the past and will be hotter in the future. But whether human civilisation can continue is uncertain.

> The world is reaching the tipping point beyond which climate change may become irreversible. If this happens, we risk denying present and future generations the right to a healthy and sustainable planet—the whole of humanity stands to lose.— Kofi Annan, Former Secretary-General of UN

The cause of the climate emergency is clear: it is due to our careless release of greenhouse gases into the atmosphere. For any readers who might still believe that climate change is a 'hoax', as a number of politicians continue to argue, here is a brief, if simplistic, overview of the science.

Earth bathes in solar radiation. So does the Moon. Earth and Moon are essentially the same distance from the Sun and, as we saw in Map 1, Earth and Moon can in many ways be considered a double planet. But the two parts of the duo possess different average temperatures. Earth is temperate, the Moon is not, and the reason for the difference is simple: Earth has an atmosphere, the Moon does not. When short-wavelength radiation from the Sun warms an object's surface then the object will re-radiate at a longer wavelength. In the case of the Moon, that re-radiation heads straight back out into space; the Moon stays cold. In the case of Earth, some of the re-radiation is absorbed by the atmosphere; Earth's atmosphere thus helps keep our planet warm. In the absence of an atmosphere Earth's average surface temperature would be about the same as the inside of a deep freeze.

The atmosphere works its wonder in large part because it contains greenhouse gases: substances that are transparent to some wavelengths of incident solar radiation (so the Sun's warming rays can reach Earth's surface) but opaque to some of the wavelengths that Earth re-radiates (so the warming effect is retained rather than lost to space). The atmosphere contains several greenhouse gases, with the most important being nitrous oxide (N_2O), methane (CH_4), and carbon dioxide (CO_2). The problem of climate change arises because humanity has unwittingly embarked upon a worrying experiment: we are changing the composition of the atmosphere, on a rapid timescale, by belching billions of tonnes of greenhouse gases into the atmosphere. Scientists long ago predicted that the outcome of the experiment would be an increase in average global temperature. And—sure enough—observations show the world is warming.

> It's not that the world hasn't had more carbon dioxide, it's not that the world hasn't been warmer. The problem is the speed at which things are changing. We are inducing a sixth mass extinction event kind of by accident and we don't want to be the 'extinctee'.—Bill Nye, 'The Science Guy'

© The Author(s), under exclusive license to Springer Nature Switzerland AG 2023
S. Webb, *Around the World in 80 Ways*, https://doi.org/10.1007/978-3-031-02440-5_4

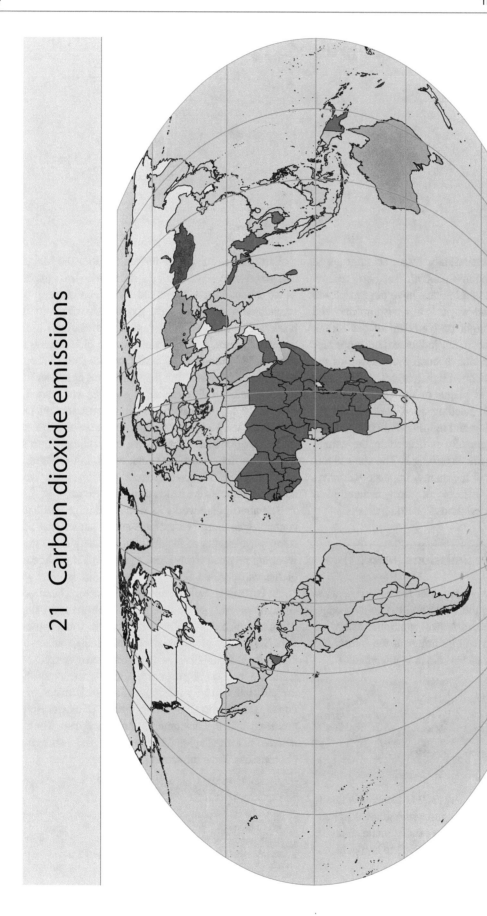

21 Carbon dioxide emissions

Carbon dioxide
emissions
(tonnes per capita)

< 1

1 – 5

5 – 10

10 – 15

15 – 20

20 – 40

No data

Map 21

Humanity began its inadvertent experiment in planetary engineering with the Industrial Revolution. Starting in the UK in the eighteenth century humans developed the habit of finding reserves of coal, gas, and oil—substances that represent frozen sunlight, energy trapped by living organisms over tens of millions of years—and then digging them up and burning them. In the ground, those reserves sequestered billions of tons of CO_2; in the atmosphere, the CO_2 warms our planet. The northern hemisphere is now about 1.4 °C warmer than in pre-Industrial times. The UN member states have set a target of limiting the average warming to no more than 2 °C above pre-Industrial temperatures, because the impact of higher temperatures is dire: sea-level rise (Map 31); disruption to water systems; changing patterns of crop growth; more droughts, floods, storms, and heatwaves. To stay within that target range we need to reduce our CO_2 emissions but over the past three decades, apart from 2008/9 (when a global financial crisis dampened economic activity) and 2019/20 (when a global pandemic dampened economic activity), each year has seen humans pump *more* CO_2 into the atmosphere than the previous year. A rise of no more than 2 °C? Good luck with that.

Although the first industrial-scale emissions of CO_2 came from the UK, many European countries quickly followed, as did North America. Other parts of the globe took some time before they began contributing significantly, but in recent decades China has become a major source of CO_2 emissions. Cumulatively, since the Industrial Revolution, the USA has emitted most CO_2: it has been responsible for more than 20% of all emissions. Countries from western Europe, combined, have been responsible for a slightly smaller total of cumulative emissions. China accounts for over 11% of cumulative emissions. Brazil (4.5% of global cumulative CO_2 emissions) and Indonesia (4.1%) are major culprits, but in the case of these two nations the damage has been caused less by the use of fossil fuels and more by emissions from their industrial-scale deforestation activities (see Map 8).

Those cumulative emissions are history. Since we face a catastrophic short-term future it makes sense to look at the CO_2 emissions happening *now*. We might then be better able to make choices that limit the damage of climate change. (We should remember, though, that the current wealth of the USA, Europe, and China rests on those cumulative emissions.)

In 2020, the latest year for which at the time of writing we have complete data, emission numbers were affected by the Covid-19 pandemic. Curbs on economic and social activity in response to the pandemic caused a 12.9% reduction in emissions for the USA, an 8% reduction in emissions for India, and a 7.7% reduction for the combined UK and European Union figure. China, however, saw only a 1.4% reduction. This meant China was by far the world's largest emitter of CO_2 in 2020. China has the world's largest population, of course, so in many ways that is not surprising. The picture changes when we look at the average number of tonnes of CO_2 emitted *per person*.

The north–south divide, with a few obvious exceptions, could hardly be clearer. Each year, most countries in Africa emit less than a tonne of CO_2 per person per year. The country with the least per capita emission is the Democratic Republic of the Congo: in 2020 each citizen emitted on average just 0.0295 tonnes of CO_2. Compare this with biggest per capita emitter: Qatar. The average Qatari emitted 1300 times more CO_2 than the average Congolese. The release of CO_2 makes Qatar rich, provides a citizens with a high standard of living (see Map 75), enables Qataris to *do* things—such as plant a forest in the desert, as discussed in Map 8. (But those trees won't absorb the amount of CO_2 emitted by Qatar.)

The citizens of Mongolia and Trinidad and Tobago, perhaps surprisingly, also have high per-capita emissions. See Fig. 24. In the case of Mongolia, per-capita emissions are high because its large reserves of domestic coal are used to supply more than 90% of its energy demand. In the case of Trinidad and Tobago, emissions come primarily from petrochemical production, power generation, and flaring. Less surprising is the fact that the oil-rich states of Bahrain, Brunei, and Kuwait also feature highly on this map.

One rung down from these super-emitters, Saudi Arabia, Kazakhstan, Australia, United Arab Emirates, Canada, and USA all emitted over 14 tonnes of CO_2 per head of population. The figure was more than that for any European country. Luxembourg (13.45 tonnes) was Europe's biggest per capita emitter; Malta (3.29 tonnes) was the EU's smallest per capita emitter; most EU countries emit in the range 5–10 tonnes per capita.

These figures highlight our dilemma. The emission of CO_2 is a byproduct of activities that define civilisation: about half of global emissions come from electricity and heat production; manufacturing and construction cause about one fifth of global emissions, as does transport; residential, commercial, and public services generate about a tenth of global emissions. We want all these activities to continue—but we need them to happen with a figure for net per capita CO_2 emission that is less than that of DR Congo! Even that might not be enough to ward off the worst effects of climate change; in addition to civilisation going cold turkey on its carbon addiction we might need to start pulling *existing* CO_2 out of the atmosphere. All this requires an unprecedented change in outlook from countries in the northern hemisphere. We look in vain to signs of that level of change taking place. As I write this, the 'anthropause' due to Covid-19 is ending.

Carbon dioxide is not the only greenhouse gas. Over a century timescale, one tonne of CH_4 has 28 times the global warming impact of one tonne of CO_2; one tonne of N_2O has 265 times the impact. Although CH_4 and N_2O are emitted in

Fig. 24 Left: an illegal coal mine in Mongolia. Many people have little choice but to risk their lives digging for coal. Each year, several miners are killed. Mongolia meets much of its energy needs through coal burning. Coal is the single biggest contributor to anthropogenic climate change. (Al Jazeera English, CC BY-SA 2.0) Right: platforms to extract gas from the Cassia field off Trinidad's coast. (Chanilim714, CC BY 3.0)

smaller amounts than CO_2, these gases still contribute to climate change. Again, northern hemisphere countries are responsible for the bulk of these emissions: methods of agriculture that help satisfy a western diet are responsible for a significant fraction of these emissions.

Is it likely that the rich, northern countries will change in time?

22 My Short-Haul Carbon Footprint (Map 22)

This map illustrates my hypocrisy.

During 2019, before Covid-19 put a stop to conferences, I was lucky enough to be invited to talk at three events: one in Scotland (where I saw the Strathmore meteorite; see Map 2) and two in Germany. In each case I took a short-haul flight from my nearest airport, Southampton. Online calculators tell me the size of the CO_2-equivalent emissions I can attribute personally to those flights: 0.42 tonnes.

In those three flights I was responsible for 17.5 times more CO_2 emission than the average citizen of the Democratic Republic of the Congo produces in one year. I was, indeed, responsible for more emission than the yearly per capita emission of many African countries. (See Map 21 for more information about yearly per capita CO_2 emissions.) I could make my point more strongly if I considered the previous year, 2018. In just one long-haul return journey from London to Vancouver I was responsible for the equivalent of 2.7 tonnes of CO_2 emission, an amount that surpasses the yearly emission figure for 2.8 billion people.

Many African nations, whose emissions I so blithely outpaced, rank among the world's least happy (see Map 50). A country's citizens might report unhappiness for many reasons, but surely a sense of inequality might be a major contributory factor. Citizens of most African countries cannot do what they see the citizens of many other countries do: burn fossil fuel with abandon in order to fly, consume produce from around the world, and enjoy the other trappings of western civilisation.

This inequity provoked within me vague feelings of shame, anger, and guilt. In honesty, however, those feelings were insufficiently strong for me to forego the convenience of flying to events of professional interest. I was able to persuade myself that those events were important; that I could offset my carbon use; that there was little I could do personally to change the political and economic order. In any case, my journeys in 2019 contributed just a few thousand of the several *trillions* of passenger-kilometres clocked up on airplanes.

But I'm a hypocrite.

A world that is warming due to the West's burning of fossil fuels threatens the existence of many people who do not themselves emit much carbon. I was able to disregard that. But harbingers of climate change chaos—fires, droughts, floods—show how the emergency threatens *everyone*. My own family included. Yet I still chose to fly.

> Dealing with global warming doesn't mean we have all got to suddenly stop breathing. Dealing with global warming means that we have to stop waste. And if you travel for no reason whatsoever, that is a waste.—Sir David Attenborough

22   My short-haul carbon footprint

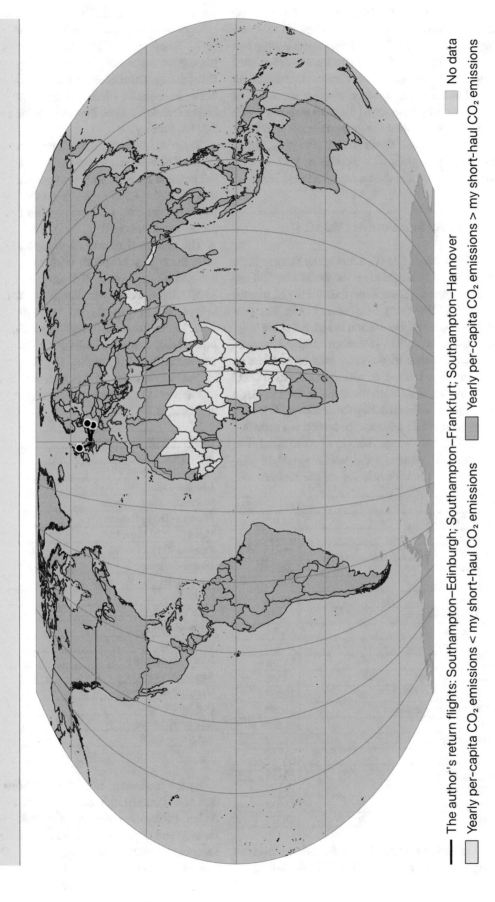

— The author's return flights: Southampton–Edinburgh; Southampton–Frankfurt; Southampton–Hannover

  Yearly per-capita CO₂ emissions < my short-haul CO₂ emissions

  Yearly per-capita CO₂ emissions > my short-haul CO₂ emissions

  No data

Map 22

In 2020, of course, the world changed: the global response to the Covid-19 pandemic shut down aviation. In this new world, though, academic conferences continued: people met online instead of in person. Online conferences offer some benefits and they present some drawbacks. Whatever academics feel about them, however, a *major* advantage of online conferences is that they avoid the costs associated with flying. None of us know what the post-pandemic world will look like when it comes to air travel. Probably, when restrictions ease, jet engines will once more start to roar. For my part, I now view flying as a habit I need to kick.

23 Busiest Airline Routes (Map 23)

The Official Airline Guide (OAG), a firm specialising in flight travel data, publishes information on the world's busiest airline routes. I am interested here in the data for the 12 months prior to February 2019—that's BC: Before Covid. The pandemic shut down much of the aviation industry, but the long-term effect of that disruption is unclear. Many indicators suggest people will return to the skies once restrictions on international travel are relaxed, in which case the final OAG report of the pre-pandemic era provides a better long-term picture than a report published during the pandemic. With that understanding we can ask: what are the busiest *international* airline routes? (I define 'busiest' here in terms of number of flights in a year. An equally valid metric would involve passenger numbers.) The table below gives the answer.

International route	No. of flights
Kuala Lumpur–Singapore	30,187
Hong Kong–Taipei	28,447
Jakarta–Singapore	27,046
Hong Kong–Shanghai	20,678
Jakarta–Kuala Lumpur	19,741
Seoul Incheon–Osaka	19,711
New York LaGuardia–Toronto	17,038
Hong Kong–Seoul Incheon	15,770
Bangkok–Singapore	14,698
Dubai–Kuwait	14,581
Bangkok–Hong Kong	14,556
Hong Kong–Beijing	14,537
New York JFK–London Heathrow	14,195
Tokyo Narita–Taipei	13,902
Dublin–London Heathrow	13,855
Osaka–Shanghai	13,708
Hong Kong–Singapore	13,654
Chicago O'Hare–Toronto	13,503
Seoul Incheon–Tokyo Narita	13,517
Osaka–Taipei	13,325

A network of routes in the Asia–Pacific region jumps out from the map. The region accounts for 15 of the 20 busiest

international journeys. The busiest route of all links Kuala Lumpur with Singapore (see Fig. 25): eight airlines fly a total of 82 times per day and between them carry 4.1 million passengers on a 300 km trip lasting about 55 min. The Hong Kong–Tapei route is less busy in terms of flights but busier in terms of passenger numbers: in the last full year before lockdown more than 6.7 million passengers made the 800 km journey.

These Asia–Pacific international routes are all short-haul flights. The busiest international *long-haul* route (defined as a journey that exceeds 4000 km) was that between New York and London: 14,195 flights made the 5585 km trip over the Atlantic Ocean on what was at that time the world's most profitable air passage.

The London Heathrow–New York JFK trip was not, however, the busiest long-haul route overall. In the BC era, the *domestic* route between JFK and San Francisco was busier: 15,587 flights were flown. Similarly, the route between Kuala Lumpur and Singapore is not the busiest short-haul route: in 2018/19, the domestic route between Seoul Gimpo airport to Juju Island, a popular holiday destination in South Korea, was served by 79,640 flights. This is the world's busiest route in terms of passenger numbers as well as flights.

The top 20 domestic routes, in terms of the number of flights, is as follows:

Domestic route	No. of flights
Seoul Gimpo–Jeju	79,640
Melbourne–Sydney	54,102
Mumbai–Delhi	45,188
Sao Paulo–Rio de Janeiro	39,747
Fukuoka–Tokyo Haneda	39,406
Hanoi–Ho Chi Minh City	39,291
Sapporo–Tokyo Haneda	39,271
Jakarta–Surabaya	37,762
Los Angeles–San Francisco	35,365
Jeddah–Riyadh	35,149
Cape Town–Johannesburg	33,708
Brisbane–Sydney	33,443
Cusco–Lima	32,095
Jakarta–Denpasar	31,958
Bogota–Medellin	31,279
Shanghai–Shenzhen	29,401
Beijing–Shanghai	29,233
Jakarta–Makassar	28,903
Bengaluru–Delhi	28,716
New York JFK–Los Angeles	26,286

By my reckoning, 48 different airports serve these routes and 46 different cities are the origin or destination of the flights. (Tokyo has Haneda and Narita airports; Seoul has Incheon and Gimpo.) The top 20 busiest international routes accounted for 333,102 airline flights. The top 20 busiest domestic routes accounted for 749,943 airline flights.

23   Busiest airline routes

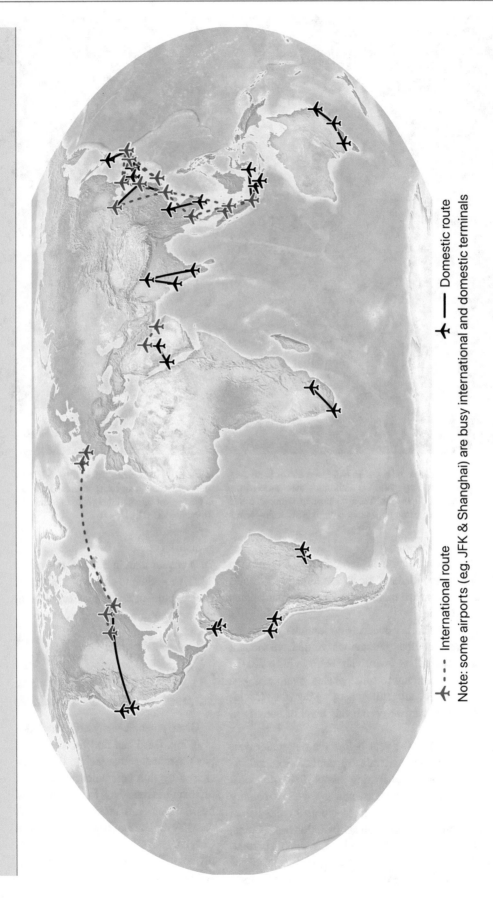

- - - - International route
——— Domestic route

✈ some airports (eg. JFK & Shanghai) are busy international and domestic terminals

Note: some airports (eg. JFK & Shanghai) are busy international and domestic terminals

Map 23

Fig. 25 Not all airports look the same. In order to cater for increasing traffic at Changi Airport, Singapore, the authorities there ordered the construction of terminal 3. The SkyTrain connecting terminal 2 and terminal 3 is visible here. More noticeable is the HSBC Rain Vortex inside the 'Jewel' area. This is the world's largest indoor waterfall. The route between Changi Airport and Kuala Lumpur International Airport is the busiest international route in terms of number of flights. (Matteo Morando, CC BY-SA 4.0)

Between them, these 40 routes carried over one million flights. Taking these 40 domestic and international routes together, the planes carried more than four *billion* passengers between those 46 cities.

On the one hand, this is a cheering notion: many of us are mobile to an extent previous generations would have thought impossible. I discussed my own flying habit in Map 22. On the other hand, the thought chills: planes pump CO_2—a greenhouse gas—into the atmosphere. After the Covid-inspired anthropause can we permit this level of activity to resume?

Worldwide, in 2017, the year before the one under consideration here, the aviation industry emitted 859 million tonnes of CO_2—thus contributing about 2% of global emissions. If the global aviation industry were a country then that year it would have ranked eighth in the list of emitters, just behind Brazil but ahead of Indonesia. (China, USA, and EU28 took gold, silver, and bronze in this particular race.) In 2020, had coronavirus not kept us home, analysts suggest the emissions from aviation would have been almost 70% greater than in 2005. If normal service resumes then by 2050, according to some forecasters, emissions from aviation could triple or even quadruple compared to the level of emissions at the start of the century.

> I think it is insane that people are gathered here to talk about the climate and they arrive here in private jets.—Greta Thunberg, World Economic Forum, Davos, January 2019

Clearly people—myself included—love the convenience of flying. But I can find little evidence people are willing to pay for schemes that seek to limit the environmental damage caused by that convenience. One simple scheme is carbon offsetting. Consider, for example, a return flight on the busiest long-haul international route, between London and New York. On that trip I would generate over 1.5 tonnes of CO_2 emissions, which is typically what I generate by heating my home for six months. If I were a passenger on that flight I could pay an extra £11.39 (about $15) to compensate for the emissions; the money would go on projects, such as tree planting, that would reduce atmospheric CO_2 by the same amount. This seems like an excellent idea. But few airlines offer a carbon offsetting program and, of those that do, passenger uptake is minimal.

An opt-in scheme perhaps goes against what we know of human nature. But the pandemic provides us with the chance of a reset. Since the aviation industry is unlikely to recover without state intervention, governments across the world are in a position to encourage a 'green' recovery. A net-zero strategy would involve improving operational efficiency (for example, by reducing the time aircraft are put in holding patterns); switching to sustainable aviation fuel (for example, using fuel produced from household waste); investing in the design of the next generation of aircraft (for example, hydrogen and electric aircraft); and managing residual emissions (for example, direct carbon capture from the atmosphere). If these approaches were successful we *might* be able to continue our flying habit without inflaming the global climate emergency. Whether governments will grasp the opportunity is debatable.

24   Renewable Electricity (Map 24)

Fossil fuels represent the frozen concentration of millions of years of sunlight. Plants in ancient swamp forests captured energy from the Sun's rays and photosynthesis produced carbohydrates. When those plants died, their remains sank to the swamp bottom and formed layers of carbohydrates. Over geological time, those layers became buried under heavy sediment deposits, and a combination of heat and pressure converted the plant remains into coal and methane. A similar process involving plankton—tiny organisms that drift in the oceans and then, when they die, fall to the ocean bed—formed, over the eons, fields of oil. When we dig up and burn these hydrocarbons—coal, gas, and oil—we release the energy of sunlight that fell upon Earth when dinosaurs were in their pomp. For many fossil fuel deposits, that sunlight warmed Earth's surface before dinosaurs even existed.

Fossil fuels are energy-dense materials: every unit mass contains a lot of energy. Burning the fuel allows us to access that energy. And when we have access to energy we can *do* things. We can transport goods and people (see Maps 22 and 23); generate electricity for light (Map 30); build imposing structures (Map 35)... indeed, we can do all the things that people in the developed world enjoy. The most powerful emperors of history could not have dreamed of the standard of living we take for granted.

There are, of course, two drawbacks to our use of these hydrocarbons.

The first problem: when we burn fossil fuels, we return stored CO_2, a greenhouse gas, to the atmosphere. Every year we release millions of years of stored carbon to the air. If we continue to alter Earth's atmospheric make-up we risk increasing Earth's average surface temperature to a value incompatible with human existence—certainly with the type of existence to which we have become accustomed.

The second problem: fossil fuels are a one-shot deal. New deposits form continuously, but natural processes require millions of years to create a seam of coal or a field of oil. Once our current reserves are gone, humanity has a long wait before it can burn new supplies of hydrocarbons. Our grandchildren will not have access to the energy-dense materials that gave rise to our technological civilisation.

In 2020, according to BP's *Statistical Review of World Energy*, just over 83% of the world's primary energy consumption came from the burning of fossil fuels (31.2% came from oil; 27.2% from coal; 24.7% from natural gas). The year 2020 was unusual, in that the Covid-19 pandemic put part of the world's economy on hold; primary energy consumption fell by 4.5%, the largest reduction in the post-war period. But as Covid-19 becomes endemic, and the world adjusts to this new reality, energy consumption will inevitably start to increase. For our own sake, and the sake of our descendants, we need to reduce our reliance on fossil fuels. And we need to do so much more quickly than we have managed to date. In recent years the share of the world's primary energy consumption from fossil fuels has declined, but nowhere near quickly enough. In 2017 the share was 85%—so in three years we saw a reduction of just two percentage points, despite almost daily news reports about the climate emergency. Modern renewables—geothermal, solar, tidal, wind; and for our purposes here we can add hydro—still represent a small fraction of total energy consumption. We need that fraction to swell. (Nuclear power offers another alternative; see Map 25. We should look at nuclear in the same way as we look at renewables: these are all technologies that, in a suitable mix, permit us to leave the remaining supplies of fossil fuels safely in the ground. In this section, however, I concentrate on the non-nuclear options.)

Let's focus on electricity generation, and the contribution of renewable sources to this critical aspect of energy policy. In the developed world we take access to electricity for granted, but not everyone is so lucky; see Map 79.

The BP *Statistical Review* gives data, for each country, of the percentage of total electricity generation contributed by renewables. Where the *Review* is missing data, the Ember website provides coverage. The map refers primarily to the year 2019, but for 69 high-consumption countries (in Europe; the Americas; Australia; and some others) we have data for 2020. The situation with electricity generation is dynamic, and some countries have diversified their energy mix, but as can clearly be seen from the map, many countries generated

24 Renewable electricity

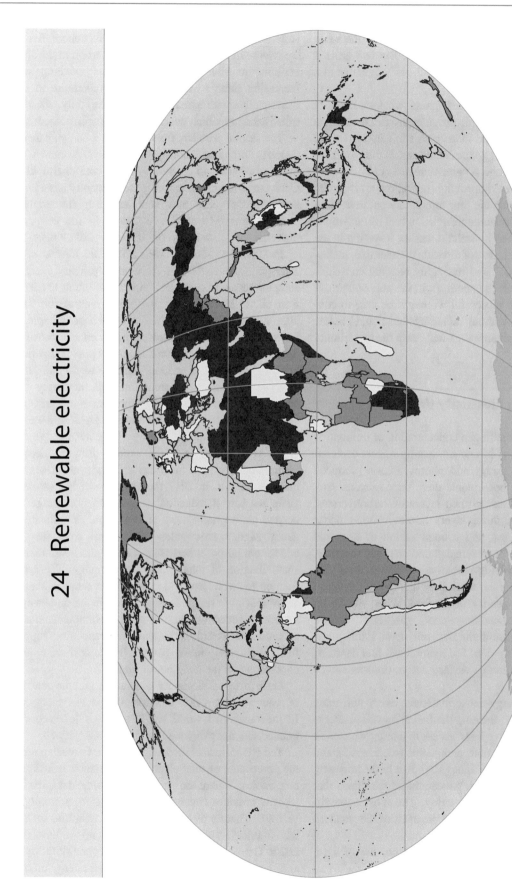

Percentage of
electricity production
generated by renewables

0 – 20
20 – 40
40 – 60
60 – 80
80 – 100
No data

Map 24

Fig. 26 In the decade 2010–2020, solar has steadily increased its share of renewable power generation. This image shows the PS20 solar power plant, which is part of the Solucar Complex in Seville, Spain. The PS20 has an array of large, moveable mirrors, called heliostats, that reflect sunlight to the top of a tower. The heat is used to turn water into steam, which drives turbines that can generate 20 MW of electricity. Also at the Solucar Complex, though not shown here, is the Solnova Solar Power Station. Solnova operates in a more traditional way and has an overall capacity of 150 MW, which makes it one of the largest solar thermal power stations in the world. (kallerna, CC BY-SA 4.0)

less than half of their electricity from renewable sources. We need to understand how our electricity is generated because initiatives we take in the belief they combat climate change, initiatives such as replacing petrol-powered cars with electric cars, may be harmful if the electricity comes burning coal, gas, or oil.

In 2020, hydropower accounted for the largest fraction (16.85%) of total electricity generation from renewable sources. The amount of energy generation from wind can vary from year to year, but in 2020 it accounted for 6.15% of total electricity generation. The share of solar power (see Fig. 26) has grown steadily, and currently accounts for 3.27% of total electricity generation. Other renewables, including geothermal and tidal, account for a further 2.72%. In total, then, less than 30% of electricity was generated from renewables; more than twice as much came from coal, gas, and oil. We have much to do if we want to avoid climate disaster.

So which countries are leading the way, and which are lagging?

Albania was among a handful of countries to generate all its electrical power from non-carbon-based sources in 2020. Albania relies on hydro. The largest energy producer in Albania is the Koman Hydroelectric Power Station: this 130-metre high dam on the Drin river drives four turbines, each of which can produce 150 MW. Paraguay likewise generated all its domestic electricity from hydro. Furthermore, Paraguay is one of the world's largest net exporters of electricity, sending its surplus to Brazil and Argentina. At the other end of the scale Guyana, for example, generated almost none of its electricity from renewables—even though it has *vast* hydro potential. And it comes as no surprise to learn that countries sitting on large reserves of fossil fuel—countries such as Libya and Oman, Saudi Arabia and Turkmenistan—generated little of their electricity using renewable sources.

Iceland still used a trace of fossil fuels, but the country's energy production came essentially all from renewables and with an interesting mix: most of its energy came from hydro but a significant fraction came from geothermal. Iceland's geology, with its high concentration of volcanoes (see Map 7), means geothermal is an attractive option for electricity generation as well as for providing hot water and domestic heating.

At first glance Costa Rica seems to be another success story: 99.8% of its electricity generation came from renewable sources with a good mix of hydro, wind, and geothermal. Even renewable power has its downsides, however. The Reventazón hydroelectric dam on Costa Rica's Caribbean coast has the capacity to power more than half a million homes, but it does so at the expense of endangering a crucial corridor for wildlife. And after seven years of planning, the country's Diquís dam was cancelled after indigenous groups protested that much of their territory would be inundated. Hydropower is cleaner than coal burning but is not without an environmental cost.

Furthermore, although some countries generate essentially all their electricity from renewable sources those same countries often find their demand for oil is *growing*. This is the case in Costa Rica: increasing numbers of Costa Ricans choose to drive cars rather than rely upon public transport, and only a small fraction of those cars are electric vehicles. When it comes to transport, then, even 'green' countries find it difficult to resist the convenience of energy-dense fossil fuels.

A country's geological situation seems, unsurprisingly, to influence its attitudes to renewables. New Zealand is a small, isolated country and so renewables are important. In 2020, more than 81% of its electricity came from renewable sources—primarily hydro and geothermal, but with wind power increasing in importance. In the same year Australia, New Zealand's much larger neighbour, generated less than 25% of its electricity from renewable sources. Australia has large reserves of fossil fuels; historically, it has been able to generate most of its power from burning coal and natural gas, and, although renewables are on the rise there, the country continues to burn fossil fuels.

Many major economies still rely on hydrocarbons to generate most of their electricity: China, Japan, USA, and several European countries generated less than a third of their electricity from renewables. The USA is particularly important in this regard because it remains the world's largest economy and has the greatest energy demands.

The USA is rich in renewables. It is the world's fourth largest producer of hydro (after China, Canada, and Brazil). Wind power is increasing in importance, and other sources of renewable energy—such as solar and geothermal—are beginning to make inroads. But can the USA make the transition to 100% renewable electricity generation? And can it stop its reliance on fossil fuels to power automobiles?

The world economy slowed down in response to the Covid-19 pandemic. As economies recover, will we see a 'green New Deal' of the type many people yearn for, or will our fossil-fuel addiction prove too hard to kick? The fate of human civilisation rests on the answer.

25   Nuclear Power Generation (Map 25)

As we saw in Map 24, many countries generate significant amounts of energy from renewable sources—hydro in particular, but also geothermal, solar, wind, and so on. Technological developments, particularly in the fields of energy transmission and storage, are likely to hasten this move to renewables. But by the middle years of this century Earth's population could reach 10 billion, and all those people will want energy for heating and cooling, cooking and lighting, entertainment and transport. Is it plausible that, over the next three decades or so, energy generation from renewable sources can keep pace with the demands of a population

swollen to 10 billion? If not, where will that energy come from?

At present we get about 80% of our energy from fossil fuels. If a population of 10 billion people is continuing to burn coal, gas, and oil to meet its energy needs then the outlook for civilisation is grim. Not only are fossil fuels a finite resource (if energy consumption continues to grow then our reserves are measured in decades) whatever remains should be left in the ground to help slow climate change. We seem to be stuck between the proverbial rock and hard place. Easy access to energy sources makes civilisation possible, and without it our way of life will perish. But continued use of our preferred energy sources will destroy our civilisation.

So what should we do?

This is a depressing dilemma. We are lucky, then, that we possess a solution. Nuclear fission reactors generate reliable, round-the-clock, carbon-free electricity; they don't belch out the particulate matter and other nasties that clog people's lungs; and they could easily scale and work with renewables to plug the energy gap created by a switch away from fossil fuels. What's not to love?

Well ... a lot, apparently. The amount of energy the world gets from nuclear is *falling*, not rising. While the global amount of electricity generated from renewables increased by 5–6% over the ten-year period 2005–2015, the amount generated from nuclear decreased by the same amount.

According to the International Atomic Energy Authority (IAEA), in 2019, the most recent year for which I have published data, only 31 countries operated nuclear power stations. In total, those power stations generated about one tenth of the world's electricity. Four countries (Bangladesh, Belarus, Turkey, and United Arab Emirates) don't currently generate nuclear power but have nuclear stations under construction. Offsetting this, some countries currently operate nuclear power stations but intend to phase them out—Germany plans to phase out nuclear by 2022, Switzerland by 2031, and several other countries are pondering whether to follow. Japan has more nuclear reactors than any country apart from the United States, but as of the time of writing only nine of its 42 reactors are in operation. As the map shows, in 2019 only three countries—Slovakia, Ukraine, and France—generated more than half of their electricity from nuclear power stations. Far from providing a quick solution to the looming energy crisis, nuclear appears to be a technology that is at best standing still rather than making progress. How did we get here?

Compared to many renewable technologies, nuclear power generation has a long history. The world's first commercial nuclear power station—Calder Hall at Windscale, on England's northwest coast; see Fig. 27—came online in 1956. (The station operated for 47 years, almost three decades longer than its design lifetime, but produced

25   Nuclear power generation

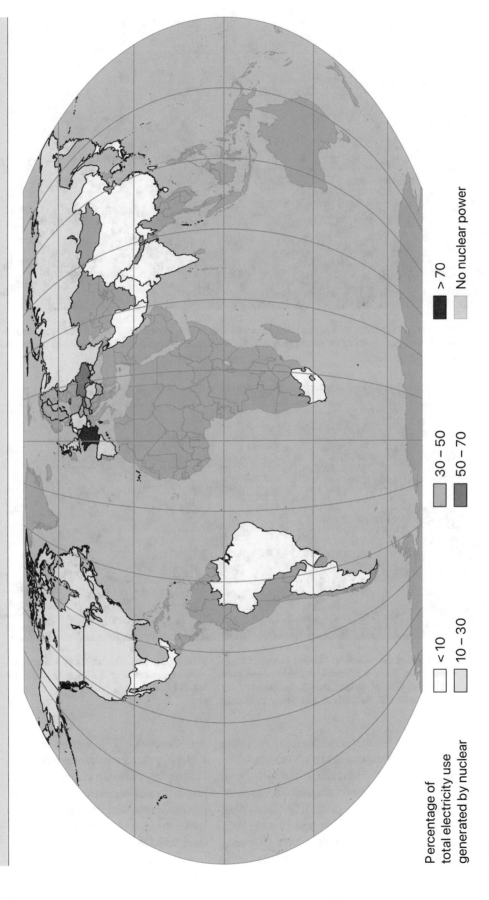

Percentage of
total electricity use
generated by nuclear

<10

10 – 30

30 – 50

50 – 70

> 70

No nuclear power

Map 25

Fig. 27 Calder Hall, soon after opening, with its two reactors, two cooling towers, and turbine house. A further two reactors and two cooling towers were later added. Modern reactors generate far more power than Calder Hall. (Public sector information licensed under OGL v3.0)

Fig. 28 Reactors 1 and 2 of the Shin-Kori Nuclear Power Plant in Korea. Since this photo was taken, a further four reactors have been added to the plant. At the time of writing, this is the largest fully operational station of its type and the tenth largest power producing plant of any kind. Operations at a larger nuclear power plant in Japan were suspended after the Fukushima incident. (Korea Kori NPP, CC BY-SA 2.0)

relatively little electricity for the domestic grid. Its prime function was to produce weapons-grade plutonium.) The Calder Hall station had two 'Generation I' reactors. From about 1965–1995, the nuclear industry employed a variety of designs and technologies in its 'Generation II' reactors; most reactors currently in operation are of this type (see Fig. 28). From 1996 onwards, various improvements led to 'Generation III' reactors. From about 2030 onwards, the hope is that the 'Generation IV' reactors currently under development will be safer, more economical, and produce less waste. Underlying all fission reactors, however, is a straightforward process: a self-sustained chain reaction generates heat, which passes to a working fluid (gas or water, depending on design), which drives a turbine, which powers a generator. This is an established technology.

Given the long history of the technology, why are so many people wary of nuclear power? The main reason, of course, is a concern over safety. Accidents happen. Nuclear accidents, if they release radioactive material into the environment, have the potential to be catastrophic.

In 1990, the IAEA introduced a scale for classifying the severity of nuclear accidents. The International Nuclear Event Scale ranges from level 1 ('Anomaly'—a minor event, which perhaps breaches statutory requirements but is unlikely to have any significant effects) through to level 7 ('Major accident'—a release of radioactive material with the potential to cause widespread damage to human health and the environment). In over 65 years of commercial nuclear power generation there have been just two level-7 events: Chernobyl and Fukushima.

The Chernobyl explosion of April 1986, caused by an improperly conducted test operation, launched radiation from a reactor core into the environment. The disaster forced authorities to abandon the cities of Chernobyl and Pripyat, and impose a 30 km exclusion zone around the reactor. The release of radiation undoubtedly led to reduced life expectancy, but the number of people whose early death can be attributed to the accident remains uncertain: a United Nations study put the number at 4000; a Greenpeace study put the figure at 200,000.

The Fukushima Daiichi disaster of 2011, triggered by a tsunami that followed the Tōhoku earthquake, was much less deadly than the incident at Chernobyl. The Fukushima reactors shut down automatically after the earthquake, but water flooded the generators that powered the pumps that cooled the reactors. The accident caused three of the station's six reactors to release radioactivity. The official death tool currently stands at one from radiation and 2202 from the subsequent evacuation.

Although we can never know for certain the number of people who died as a result of these two disasters, we can be sure that the figure is measured in the thousands. This was sufficient for the German government to hasten its plans to phase out nuclear power. Other governments will follow the German lead. But the calculus in these situations is not so simple.

If Chernobyl and Fukushima had never been built then the power they generated would presumably have been provided by the burning of fossil fuels. (The alternative—making do without that energy—would not have been an option: people's appetite for electricity is insatiable.) But getting energy from fossil fuels is *dangerous*: every year thousands die in accidents during the extraction processes, thousands more from the pollution spewed out by burning oil and coal. In fact, for each unit of energy generated, nuclear power has been the cause of fewer accidental deaths than oil, gas, and coal. Nuclear is even safer than hydropower: when dams collapse, thousands die.

Nuclear power undoubtedly has drawbacks. But we are in the midst of a climate change emergency. If we continue to use fossil fuels then we won't be counting thousands of deaths from explosions and fires and pollution, but many *millions* of deaths from drought and famine and heat. Given the emergency, can we afford to ignore nuclear?

A different nuclear technology has the potential to provide an almost inexhaustible supply of energy. The reactors discussed above, from Calder Hall through to the Generation IV technology currently being developed, all rely on *fission*: large atomic nuclei split into smaller nuclei, with the release of energy (which we can harvest) and neutrons (which cause other nuclei to split and the reaction to continue). *Fusion*, the process that causes stars to shine, has small atomic nuclei fuse together to create larger nuclei with an associated release of energy. The raw material for fusion comes from seawater and, on a mass basis, fusion generates more power than fission. Fusion could thus be the answer to our energy problems. Unfortunately, a commercial fusion reactor remains tantalisingly out of reach. With current technology, engineers must put more energy *into* a fusion reactor than they get *out*. Since the 1970s, a joke goes that commercial fusion power is about forty years away—and it always will be. The climate change emergency no longer allows us the luxury of four decades of design and development of a complex technology. Why not use a proven technology, fission, to fill the immediate gap?

26   Trash (Map 26)

I am just about old enough to remember a time before throwaway societies became the norm in the developed world. Consumers bought things in the quantity they needed. To discard good food was considered a sin; when clothes began to wear thin one patched them; broken tools got fixed. Nowadays, at least in some countries, calories are so plentiful it seems wrong to ingest them all; clothes are bought to last a season; tools with electronic parts are often impossible to mend. So we throw things away.

> Eat it up, wear it out, make it do, or do without.—Calvin Coolidge, 30th president of the United States

Municipal solid waste—trash, in other words—is the stuff we modern-day consumers chuck away after use: bottles and boxes; food and fridges; car tyres and computer tablets. Trash is the detritus of our everyday lives. (Our activities of course generate other types of rubbish: debris from construction; wastewater sludge; industrial waste. But that is something else again.) The 'What a Waste' project aims to collect and

26 Trash

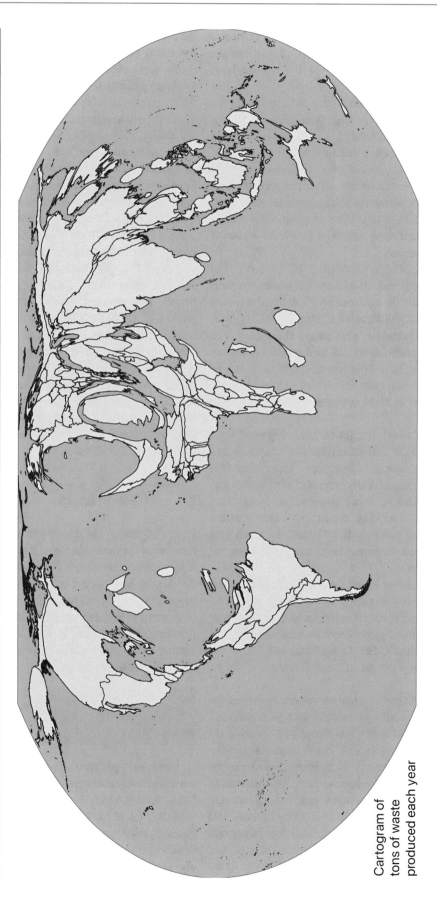

Cartogram of
tons of waste
produced each year

Map 26

publish data on all aspects of the management of solid waste in countries across the world. Although the quality and breadth of data inevitably varies from country to country, the project's database is the most comprehensive available at the global level. The project collates information on solid waste management from almost all countries and from 360 cities. One depressing statistic from the project is that the global amount of municipal solid waste—all that stuff we throw away after use—is expected to increase from about 2 billion tons today to 3.4 billion tons by 2050.

What happens to all that waste? Well, globally, a third of total waste is openly dumped; about 37% goes into some form of landfill; 11% is incinerated; and only a fifth is recovered through recycling and composting. If these proportions are maintained when the world throws away 3.4 billion tons of waste then, especially for countries that use open dumping, the environmental and health implications will be dire.

In this map—a cartogram rather than a choropleth—the area of each country is scaled depending on how much trash it generates each year. The USA, China, India, and Brazil maintain their size: they generate the most trash. But some other countries with a large land area, such as Australia, Canada, and Russia, shrink: they generate less trash. When the world is scaled in this way, the continent of Africa no longer dominates. On the other hand, some islands attached to a mother country can appear artificially large.

A cartogram only works when the number being plotted involves an overall total. Consider trash on a per capita basis and the situation alters. Iceland, a tiny island with a rich and highly consumerist population, generates 1587 kg of trash per person per year—the most in the world. Although Iceland has a reputation for 'green' credentials (see Map 24) its people's consumption patterns, in a land where most things must be imported and exported, are extravagant. A 2016 study undertaken by the Environment Agency of Iceland reinforces this: on average, each resident of the island throws away 23 kg of consumable food and 39 kg of spoiled food every year. The island's restaurants throw away 40,000 tons of food.

In Europe, three countries I believed were 'green'—Denmark, Germany, and Switzerland—produce more trash per person than France, Italy, or the UK. And the USA and China, those two biggest producers of trash in absolute terms? Well in one year an American citizen produces 810 kg of trash, a Chinese citizen just 153 kg.

27   Plastic Waste (Map 27)

The 'What a Waste' project referenced in Map 26 records details on the type of trash we produce—the percentage, by weight, of municipal solid waste in the form of glass, food

and organic, metal, and so on. Plastic is a particularly troubling type of trash.

Our civilisation produces mountains of plastic: since the mid-twentieth century our factories have manufactured 6.3 billion tons of the stuff. About 21% of this has been recycled or incinerated, which means about 5 billion tons of plastic has not been dealt with. Most plastics, however, resist the natural processes that break down other substances. Much of that 5 billion tons of plastic therefore endures—polluting land, rivers, and oceans. And the problems caused by plastic pollution are becoming increasingly evident. Plastic debris kills wildlife in a number of ways. Its effect on humans is still being researched.

Palau, a country of about 340 Pacific islands, has a particular problem: almost one third of its municipal solid waste is in the form of plastic. The Federated States of Micronesia, which along with Palau form the Caroline Islands, is almost as bad at generating plastic waste. These beautiful Pacific islands, however, do not contribute much waste in absolute terms: their combined population is the same as a smallish town in England. Larger countries pose a bigger threat. In Kazakhstan and South Korea, for example, a quarter of municipal solid waste is plastic; in Great Britain just over one fifth of trash is plastic.

Industry has started to experiment with alternative plastics that break down on a timescale of weeks, rather than centuries, as is the case with traditional plastics. Two main types of alternative plastic are in use: *bioplastics* are made from organic materials such as corn starch rather than petroleum, while *biodegradable plastics* are made from petroleum-based plastics combined with an additive that makes them degrade quickly. The two terms are often used interchangeably, but chemically the two types of plastic are quite different. So—has this solved the problem? Unfortunately, no.

Consider bioplastics. (Note that an ethical question arises over the use of these substances. The raw material for bioplastic is a crop that requires agricultural land, energy, and water to grow. Is it ethical to use these resources for making plastic instead of food? Here, though, I concentrate on its more direct environmental impact.) Bioplastics *can* be broken down in an environmentally friendly way—but only if they are collected and composted in specially designed composting stations. Many countries lack those industrial facilities. The typical fate of many bioplastics is to end up in landfills, with insufficient oxygen to allow the substance to break down. Bioplastics can then last for decades, even centuries, and release methane—a more potent greenhouse gas than carbon dioxide.

Or consider biodegradable plastics. The lazy consumer, operating under the natural assumption that a biodegradable bottle will, well, degrade, might feel no guilt at casually throwing the bottle away. But if that piece of biodegradable

27 Plastic waste

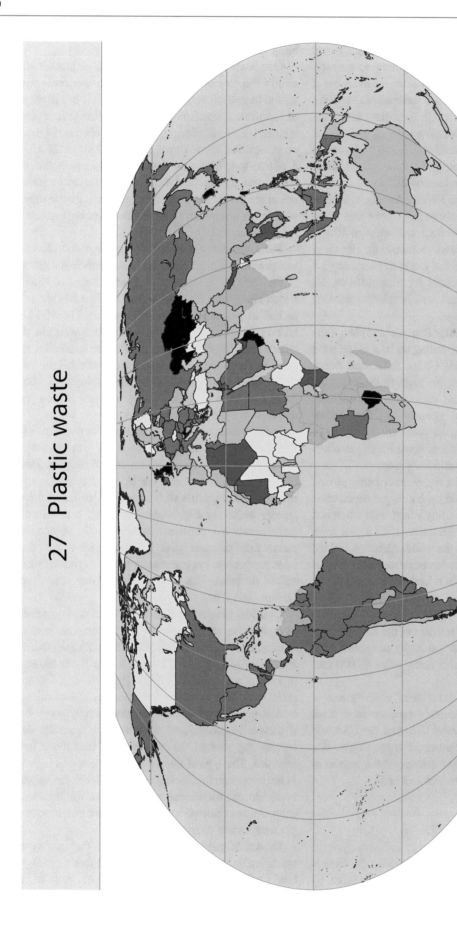

Percentage of trash
(by weight)
that is plastic waste

< 5

5 – 10

10 – 15

15 – 20

> 20

No data

Map 27

plastic reaches the ocean it will not quickly break down. It will harm marine life and eco-systems. And where biodegradable plastic *does* degrade, in many cases it leaves behind toxic microplastics.

A further problem with both bioplastic and biodegradable plastic is that they often contaminate the conventional plastic waste stream, which hinders recycling.

Technologists will surely develop an environmentally friendly, economically viable alternative to plastic. Until they do, however, the best way to address the problem of plastic pollution is to remove plastic wherever possible from products and packaging. Where plastic can not be removed, its use should be reduced. And, wherever possible, plastic should be reused.

28 Pesticide Use (Map 28)

People have used pesticides for millennia. The farmers of ancient Sumer, for example, dusted their crops with sulphur in order to killed insects. It is only in recent decades, however, that have farmers come to *depend* on the deployment of herbicides, insecticides, fungicides, and their array of other cides. Even with such an arsenal behind them, they face a difficult task: the damage caused by pests reduces annual global crop yields by up to 40%. Without weedkillers and other chemical weapons farmers would struggle to feed a world population of more than seven billion. On the other hand, pesticides can lead to health problems in humans. They also create broader environmental problems: the indiscriminate use of pesticides, along with factors such as light pollution (as seen in Map 30), are causing an alarming decline in the insect population. Little wonder, then, that the Food and Agriculture Organization (FAO), a United Nations agency whose goal is to achieve food security for all, is interested in understanding the use of pesticides. (The FAO define pesticides as "insecticides, fungicides, herbicides, disinfectants and any substance or mixture of substances intended for preventing, destroying or controlling any pest".)

So—how much pesticide is being used around the world? Well, the FAO collect data on the total amount of active ingredients in all types of pesticide put on land. The range of use is vast.

At one end of the scale lies the Maldives, which uses 52.6 kg of pesticide per hectare of cropland. The pesticides are primarily used for mosquito control (see Map 56) rather than crop protection, but not only mosquitos are affected: in recent years the number of insect species in the archipelago has dropped by two thirds. Pesticide use in the Maldives is more than twice that of the next most intensive user, Trinidad and Tobago, which applies 24.9 kg per hectare. The other places using more than 20 kg per hectare are Costa Rica, Bahamas, and Barbados. In Costa Rica, pesticides are used on important export crops such as banana and coffee. The practice has led to health problems in humans, ranging from acute poisoning to chronic disease, and environmental contamination through spray drift and run-off.

At the other end of the scale many African and Asian countries use only trace amounts of pesticide. In Europe, Nordic countries provide a good example of how to manage pesticides. Starting in the 1980s, Denmark and Sweden began to put in place policies to reduce the amount of agricultural pesticide in use. The intention was not to eliminate pesticides entirely, since often they are necessary, but to use them in a more targeted manner. Decades later, the result is that Sweden uses almost seven times less pesticide per hectare than Germany; Denmark uses more than seven times less pesticide per hectare than the Netherlands. The Nordic countries are now researching further improvements, such as implementing new technologies for pesticide application and increasing the number of buffer zones so that water quality and biodiversity are not harmed.

The 1960s was the decade of the Third Agricultural Revolution: Norman Borlaug introduced new, higher-yielding varieties of wheat and in doing so saved a billion people from starvation; farms became increasingly mechanised; the use of chemical fertilisers became widespread. And pesticide use went up exponentially. Although the global population *increased* (it has more than doubled since I was born) food prices *decreased* because farming yields improved. Science prevented a humanitarian disaster. By 2050 the world population will be almost 10 billion. Can Science feed those mouths without using yet more pesticides, with all the problems that might bring?

28 Pesticide use

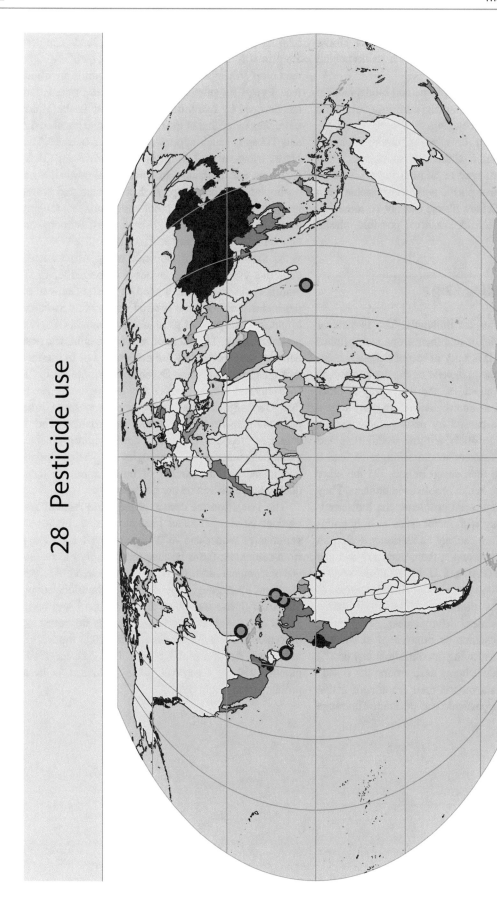

Pesticide use:
kg per hectare
of cropland

<1	2 – 6
1 – 2	6 – 10
	10 – 20
	> 20
	No data

Map 28

29   Mammal Species Under Threat (Map 29)

In 1948, UNESCO helped found the International Union for the Protection of Nature. The organisation's name has changed slightly over the years—it is now the International Union for Conservation of Nature and Natural Resources (IUCN)—but it still assists, encourages, and influences societies everywhere to conserve nature and use natural resources in a sustainable way. One way it does this is by gathering data on the state of the natural world, with its most important publication being the IUCN Red List of Threatened Species. Since its establishment in 1964, the Red List has become the most comprehensive measure of the conservation status of plant, fungi, and animal species.

At the time of writing, the Red List contains an assessment of the conservation status of 98,500 species across the living world. Species fall into one of nine classes: *not evaluated* (although 98,500 species might seem a lot, countless more could be assessed and the IUCN hopes to soon double the current number); *data deficient* (this applies when we have insufficient understanding of the population or distribution of a species); *least concern* (such species are not a focus for conservation activities); *near threatened* (species not currently threatened with extinction but might soon become threatened); *vulnerable* (species which, usually through habitat loss, are likely to become endangered unless circumstances change); *endangered* (species that face a high risk of extinction in the wild); *critically endangered* (species that face an extremely high risk of extinction in the wild); *extinct in the wild* (species that survive only in captivity or outside their native range); and *extinct* (we have no reasonable chance of seeing them again).

A list such as this is dynamic. For example, new information on population size might come to light or biologists might change the taxonomic classification of a species; conservation measures might alleviate problems or environmental threats might increase them—developments can be positive or negative. Overall, though, the news is bad. Of the 98,500 species on the Red List more than 27,000 are threatened with extinction. Any decline in biodiversity should make us uncomfortable: lose pollinators and we face a threat to our food supply; lose fungi and we are deprived of sources of possible new medicines; lose flowers and we remove beauty from the world. Humanity's inherent chauvinism, however, means most of us are more interested in the extinction threat to mammals. There too the news is bad: of the mammal species assessed by the IUCN, 25% live under threat of extinction. Indonesia fares worst.

Indonesia's 18,000 islands are home to 515 mammal species; 191 of them are threatened. Habitat loss plays a major role. The world's third largest rainforest lies here (see Map 8) and, despite covering only 1% of Earth's surface area, it contains 10% of the world's classified plant species, 17% of its bird species—and 12% of its mammal species. As burning and logging shrink the Indonesian rainforest, mammalian biodiversity shrinks too. Some of the critically endangered species are iconic. The Sumatran tiger is the last remaining species of Indonesian tiger, and extinction looms. (The Balinese and Javan tigers are already extinct.) The Javan rhinoceros is down to about 60 creatures. The Sumatran and Bornean orangutans, among the most intelligent of primates, are under threat from an even more intelligent primate.

Indonesia's problems are not unique. Since Madagascar split from India, about 88 million years ago, a distinct ecosystem has evolved. Most of the mammal species on Madagascar can be found nowhere else, and the Red List tells us 121 of them are threatened. And so it goes. In Mexico, many mouse species are under threat; in India, the pygmy hog faces extinction; in Brazil, two capuchin species are endangered. As humans destroy habitats and heat the planet, more creatures will become extinct. And we, too, might join the Red List.

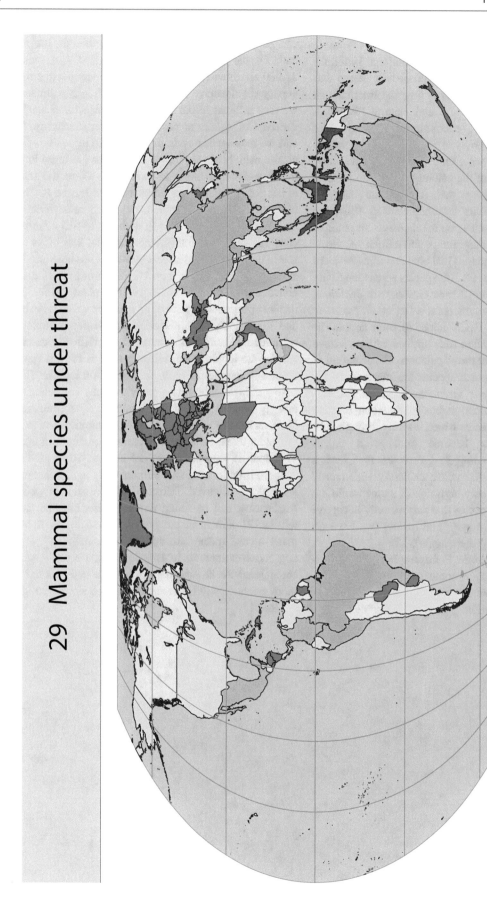

29 Mammal species under threat

Number of
mammal species
under threat

<10

10 – 25

25 – 50

50 – 125

> 125

No data

Map 29

30   Night Sky Brightness (Map 30)

The chances are high you have never seen a truly dark sky. In 2016, an international team of scientists combined high-resolution satellite data with precision brightness measurements to generate an atlas of artificial night sky brightness. The atlas showed that 99% of Americans and Europeans live under light-polluted skies. More than one third of humanity experiences a night sky that obscures the Milky Way. Although we know more about our place in the cosmos than ancient astronomers could ever have imagined, we have fewer opportunities than our ancestors to directly *observe* the universe. The thought saddens me. We have diminished a fundamental, shared human experience—looking up and marvelling at the night sky.

Before humans began polluting the night sky with light, illumination came solely from natural sources: airglow (the emission of light from various atmospheric processes); celestial objects (primarily the Moon, stars, and Milky Way); and zodiacal light (the reflection of sunlight from ice and dust particles in the plane of the solar system). Taken together, these natural sources of night sky light created a luminous intensity level of 0.000174 candela per square metre. The candela is the relevant SI unit and, to give some idea of what it means, a common household candle emits light with a luminous intensity of about 1 candela. So to get a feel for the value 0.000174 cd/m^2 imagine the light of a single candle spread over the area of a football pitch. In the absence of anthropogenic sources of light, the night sky is *dark*.

The disappearance of the dark night sky is an unwanted side effect of civilisation. We light up the interior and exterior of our houses; our roads glow beneath streetlights; advertising boards and sporting venues shine; we illuminate offices, factories, and commercial properties. But thought is seldom given to how best shield and target this lighting, or even whether we need the lighting in the first place. And, as is often the case with environmental matters, addressing one issue can cause problems elsewhere: in the UK, yellow sodium streetlights are being replaced by blue-white LED lights—lower carbon emissions but worse light pollution. In sum, photons from all these artificial sources scatter into the atmosphere, increasing the night sky luminance and creating a human-made skyglow. Even if you live far from a city, that artificial skyglow can rob you of a view of the dark night sky. Unwanted light represents one of the most pervasive forms of pollution, all the more difficult to deal with because of its insidious nature. Many of us don't even realise it's there.

A handful of Earth's inhabited places—Cocos Islands; Niue; Norfolk Island; Tokelau; and Wallis and Futuna—are so remote that essentially *everyone* there enjoys a night sky with a natural luminous intensity level. But these are small islands in the Indian and South Pacific Oceans. Of large countries, Chad fares best (79% of its population can enjoy natural night sky brightness), followed by Central African Republic (78%) and Madagascar (77%). In Europe, few citizens can experience the night sky in the way people once enjoyed it: the artificial airglow is all-pervasive. Worldwide, the situation is worst in Singapore: streetlamps light up its roads and walkways; its office buildings glow late into the night; the island's container terminals, airport, and financial district shine day and night. Not one inhabitant of Singapore can see the Milky Way: everyone suffers an artificial night sky brightness greater than 3000 cd/m^2.

Does this matter to anyone but astronomers? Well, yes. Apart from the cultural loss involved, light pollution affects the day–night rhythm governing our biological clock. The result is an increased risk for developing obesity (see Map 53), depression, and diabetes. Light and chemical pollution combines with climate change to form a triple-whammy for insects. Even an insectophobe should worry about declining insect numbers: if insects vanish, so might we.

30 Night sky brightness

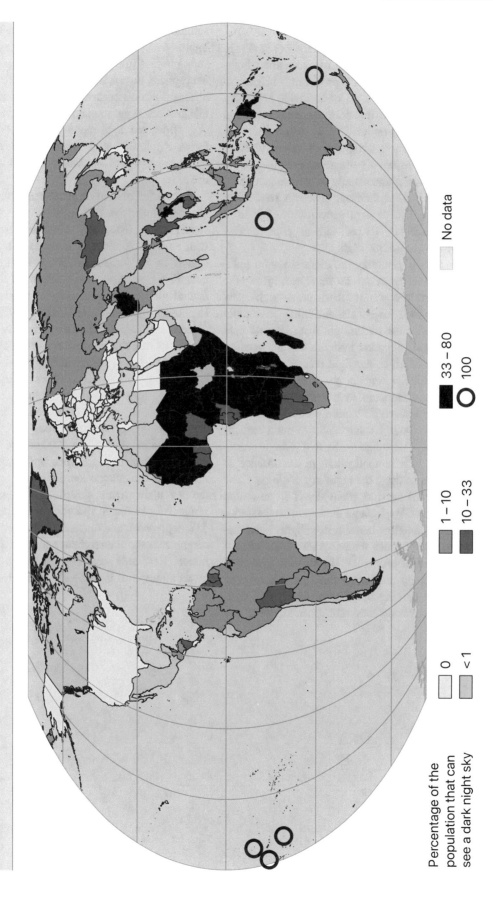

Percentage of the
population that can
see a dark night sky

0
<1
1 – 10
10 – 33
33 – 80
100
No data

Map 30

The World of the Built Environment

A stone wall stands in front of Theopetra cave in Thessaly, Greece. Some nameless masons built it 23,000 years ago, to protect the cave's inhabitants from biting cold winds (they were living at the height of the last Ice Age). The Theopetra wall may well be the oldest human construction on Earth. Since then, of course, our building skills have improved.

The Egyptians built the Great Pyramid, 146.5 m tall; modern architects have erected a skyscraper that reaches a height of over 800 m. The Romans built amphitheatres for gladiatorial games; today, we build stadia for football games. We build ports, not just to launch sea craft but to launch space craft. We build shops, and offices, and restaurants. We build electricity generating plant, and transmission grids to deliver that power to our shops and our sporting arenas, our ports and our restaurants. We build supercomputers, which in turn enable us to design taller skyscrapers and safer spaceports and more luxurious malls. In 2020, global human-made mass exceeded the overall living biomass.

Our buildings are changing the face of the planet. But this frenzy of construction has been made possible by the energy released through the burning of fossil fuels. And that in turn is warming our planet. As Earth warms so the oceans expand and many of our buildings—from the humblest of homes to buildings of such importance they populate the UNESCO World Heritage List—are at risk of sinking beneath the waves.

---

31 Homes at Risk of Rising Sea Levels (Map 31)

In 1998, Joel Cohen (a mathematical biologist) and Christopher Small (a geophysicist) initiated the field of hypsographic demography. This trochaic tetrameter is a rather convoluted way of expressing a simple idea: Cohen and Small studied the distribution of human populations with respect to altitude. The two scientists combined census estimates from 1994 with new 3D digital maps of Earth and showed that more than a third of the world's population lived less than 100 m above sea level. Most of us probably have a vague intuition that many populous cities are low-lying, and that intuition is correct. When Cohen and Small wrote their paper, 15 cities had a population of more than 10 million people, and 11 of those 15 had an elevation below 100 m: Shanghai (elevation 4 m), Rio de Janeiro (5 m), Calcutta (9 m), New York (10 m), Mumbai (14 m), Osaka (24 m), Buenos Aires (25 m), Seoul (38 m), Tokyo (40 m), Lagos (41 m), and Los Angeles (87 m). The surprise—then and now—is that most people living at an elevation below 100 m do *not* live in mega-cities. They live in areas of much lower population density. I number myself amongst them: my family lives close to the coast, at an elevation of 10 m.

Hypsographic demography, it turns out, has applications in many fields—from biomedical research to semiconductor manufacture to food production—but, from the outset, Cohen and Small were thinking about people living under the threat of coastal flooding. Planners involved in disaster preparedness need to understand how best to allocate limited resources, and the surprising result of Cohen and Small—that more people live in low-lying villages and towns than live in low-lying large cities—was an important piece of information. What I find disturbing is that these two American scientists were considering how to mitigate the increasing risk of flooding more than 20 years ago. Since then, human activity has continued to make the threat of catastrophic flooding much more likely. Although some countries are already experiencing more frequent and more severe episodes of flooding, there is no obvious indication that civilisation might change tack.

Since the start of the Industrial Revolution humans have released large quantities of CO_2 and CH_4—potent greenhouse gases—into the atmosphere. The changing atmospheric composition is having the predicted effect: our world is warming. This in turn has an effect on sea level. A warming sea occupies more volume—that's just the effect of thermal expansion—while the melting of ice sheets adds

S. Webb, *Around the World in 80 Ways*, https://doi.org/10.1007/978-3-031-02440-5_5

31 Homes at risk of rising sea levels

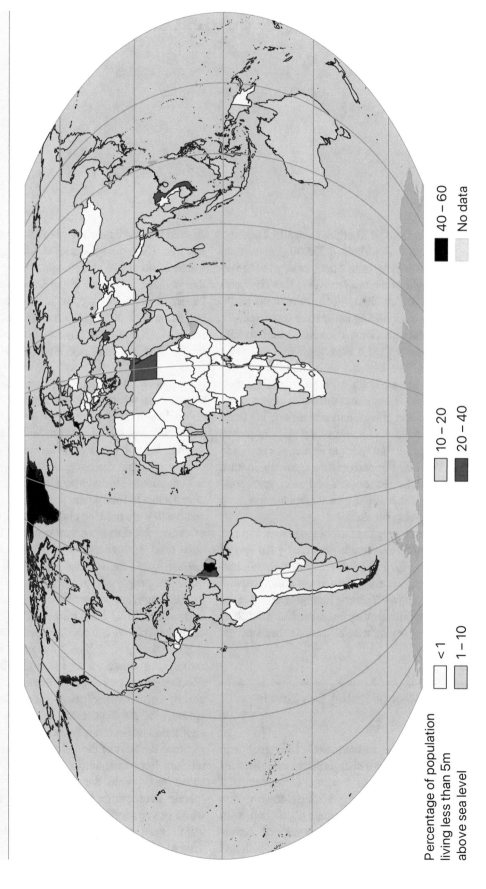

Percentage of population
living less than 5m
above sea level

☐ <1	☐ 10 – 20	■ 40 – 60
☐ 1–10	☐ 20 – 40	☐ No data

Map 31

Fig. 29 The 5-km-long Haringvlietdam, which connects the islands of Voorne-Putten and Goeree-Overflakkee by road, is part of the Delta Works—a set of barrier protection works against the North Sea. Construction of this concrete-and-steel structure began in 1957 and was completed 14 years later. (Public domain, CC0)

extra water into the oceans. The melting of ice only increases sea level if the ice currently covers land, of course; the melting of a free-floating iceberg has no affect on sea level, just as the melting of ice cubes does not cause the drink in a glass to overflow. But there are two large land-covering ice sheets on Earth: Antarctica and Greenland. Antarctica holds about 30 million cubic kilometres of ice (see Map 1); if all that ice were to melt, sea levels would rise by about 60 m. If all of Greenland's 2.85 million cubic kilometres of ice were to melt, sea levels would rise by about 7 m. Clearly, any such event would be catastrophic for human civilisation.

Although such an apocalyptic event seems unlikely to occur, a business-as-usual approach to greenhouse gas emission means rising sea levels will cause misery for hundreds of millions of people. The Intergovernmental Panel on Climate Change estimates that, over the past century, warming has caused the average global sea level to rise between 100 and 200 mm. The rate of increase, moreover, is accelerating: sea levels are rising by about 2.5 mm per year. These might appear to be tiny amounts, particularly in comparison with the sea level rises associated with the doomsday event of ice sheet collapse, but even small increases in sea level can have profound effects.

Bruce Douglas, an American oceanographer, argues that, for every 25 mm of rising sea level, erosion can cause the horizontal retreat of 2.4 m of sandy beach. The shape of our coastline could change almost before our eyes. But coastal erosion is a relatively trivial problem compared to some of the other effects. If salt water contaminates freshwater aquifers then agriculture becomes difficult and water is no longer potable. Storm surges become more dangerous so we would see more events like the North Sea flood of 1953, which killed 2551 people. Tsunamis would be even more devastating. Some nations could simply disappear under the water, a series of modern-day Atlantises.

The map in this section shows the percentage of a country's population living in areas where the elevation is below 5 m—people who are particularly threatened by rising sea levels. One of the countries under most threat is the Netherlands: 58.5% of the population lives at an elevation below 5 m. This is unsurprising: about half of the country's land is already at or below sea level. Having lived in Amsterdam for a couple of years I have a healthy respect for the engineering ability of the Dutch. They do, after all, have a centuries-long history of battling the sea. See Fig. 29. But I still remember visiting the coast and experiencing the disconcerting sight of a ship sail past on the North Sea while it was at a higher elevation than me. Can such a low-lying country survive in a world of rising sea levels?

The Dutch have already built significant flood defences. Following that North Sea flood of 1953 they began the Delta Works, a collection of dams, sluices, and storm surge barriers

Fig. 30 The Funafuti atoll of Tuvalu looks paradisical, but Tuvalu risks being submerged because of rising sea levels. (mrlins, CC BY 2.0)

that has been called one of the seven wonders of the modern world (see Map 33). The idea behind the project was to develop a theoretical framework that could underpin practical, long-term solutions to protect the country from flooding. The framework rested on several principles, including an assessment of the cost of a flood in terms of property damage, lost economic production, and human life. (For the purposes of the framework, as of 2008 the Dutch valued a human life at 2.2 million euros.) A flood simulation lab calculated the chances of flooding in different areas, and then the most cost-effective flood defences were constructed.

The Dutch have also been innovators in urban design: Watersquares provide a water store during periods of heavy rain and a community gathering space during periods of drought; green roofs manage water and keep structures cool during summer; in IJburg, they are experimenting with floating houses. So perhaps the Netherlands, a rich country with engineering expertise, can survive whatever the ocean throws at it. Other countries may be less adaptable.

In Suriname (see Map 8 for more about this country), 56.2% of the population lives at an elevation below 5 m. Agriculture is critical to the Suriname economy, but much of the country's fertile land lies in the coastal plane. If encroaching salt water makes agriculture more difficult then the people of Suriname will face a grim future. Might they be forced to migrate?

Low-lying islands such as Turks and Caicos Islands, Maldives, and Tuvalu may become uninhabitable. Turks and Caicos Islands has already experienced the effects of a category 5 hurricane; Maldives, the flattest country on Earth, will lose 77% of its land area by 2100 given mid-level scenarios for CO_2 emissions; a rising saltwater table means Tuvalu will lose many of its crops, and rising temperatures have threatened traditional fisheries; see Fig. 30.

Ten years after Cohen and Small began their research, Gordon McGranahan, Deborah Balk, and Bridget Anderson revisited their findings. The demographics and direction of travel are clear. About one in ten humans live less than 10 m above sea level. Two in three of the world's large cities (those with more than 5 million inhabitants) are in low-lying coastal areas. The period since the McGranahan paper was published has seen 8 of the 10 hottest years on record. Sea levels will continue to rise and homes will be washed away. And, on the evidence to date, there doesn't seem to be much humankind is willing to do about it.

32 Lighthouses (Map 32)

I'm a lighthouse fan. These romantic structures are often found in spectacular locations and so offer great views. (For a long while I also thought they were the best example of a

32 Lighthouses

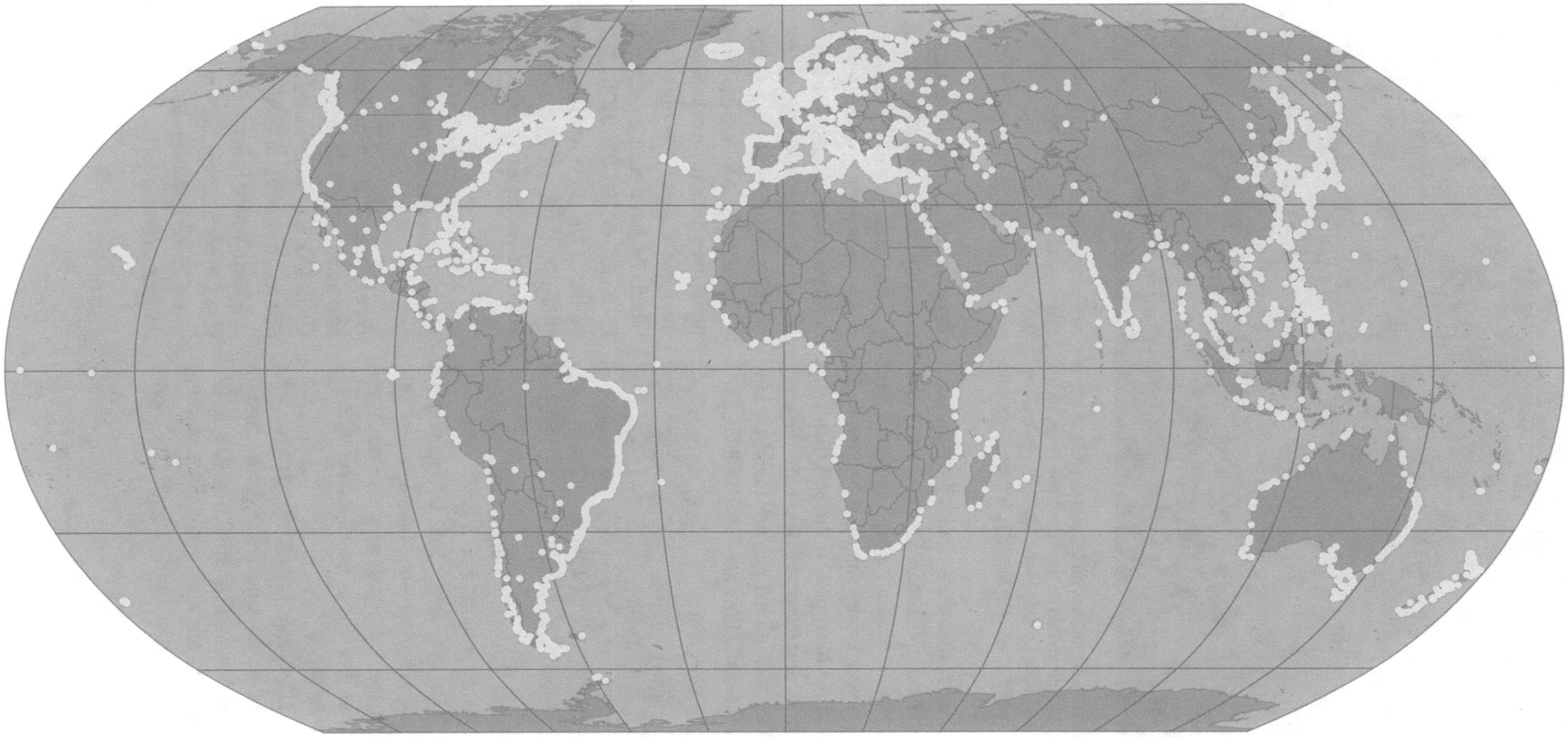

Location of a lighthouse

Map 32

Fig. 31 The French physicist Augustin-Jean Fresnel developed a lens capable of making a light beam visible to the naked eye from 30 km away. On 25 July 1823 he lit such a lens in the Cordouan lighthouse, shown here (now a world heritage site; see Map 33). Improved versions of the lens were installed in lighthouses across the world. It is often said that Fresnel's invention saved a million ships. (Gadjo Niglo, CC BY 2.0)

public good, a service the marketplace cannot provide, but it turns out that lighthouses in Victorian England were often privately owned. The owners couldn't charge shipowners for the service, of course. Instead, they sold their service to the owners or merchants of the local port. If the port didn't pay up then off went the light, and the port would have difficulty in attracting ships. Oh, well.) This got me thinking: how many lighthouses are there and how are they distributed?

> Were I a Roman Catholic, perhaps I should on this occasion vow to build a chapel to some saint; but as I am not, if I were to vow at all, it should be to build a lighthouse.—Benjamin Franklin, July 1757, after surviving a shipwreck

Russ Rowlett's *The Lighthouse Directory* has details of over 21,200 lighthouses around the world. The UK Admiralty's *Digital List of Lights* has information on 85,000 individual light structures—but that includes light ships, lit floating marks, and fog signals as well as traditional lighthouses. Rather than use these excellent resources, I decided to make my own list.

OpenStreetMap, a UK-based community project with more than 2 million volunteers around the globe, aims to provide free geographic data to anyone who wishes to use it. It's easy to query the database and extract the location of all structures that volunteers have tagged as being lighthouses. My query returned 29,672 objects, which is certainly in the right ballpark.

Plot their location and we see that this number of lighthouses is sufficient to trace the coastlines of the world's continents and islands. The lighthouses even flag the location of several major lakes and rivers. This should not come as a surprise: lighthouses have a long history.

In antiquity, a hilltop fire served as an entrance marker to a port. Over time, the markers began to function as a signal of maritime danger; see Fig. 31. The earliest lighthouse for which we have clear evidence is the Pharos of Alexandria, built around 280–250 BCE. This was the third longest surviving of the seven Ancient Wonders, finally disappearing in 1480. In 1994, archaeologists found remains of the lighthouse on Alexandria's Eastern Harbour floor.

33   World Heritage Sites (Map 33)

As of summer 2021 the United Nations Educational, Scientific and Cultural Organization, more commonly known as UNESCO, listed 1154 sites as possessing such importance—cultural, historical, scientific or natural—that their protection and preservation is in the collective interest of humanity. The idea that certain places are so valuable they must be protected began to be discussed in 1954, when Egypt's government decided to construct the Aswan High Dam. It was clear this

33   World heritage sites

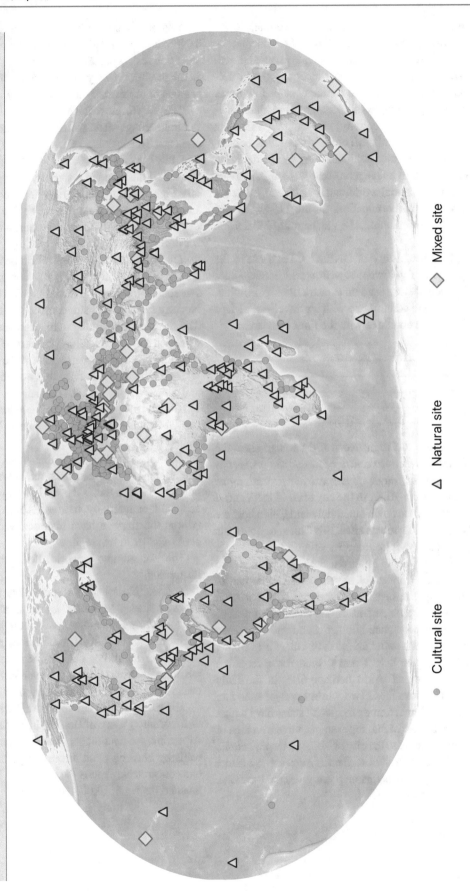

Cultural site   •

Natural site   △

Mixed site   ◇

Map 33

project would inundate areas that were home to many items of huge cultural significance, so UNESCO launched an appeal to excavate, record, and relocate these treasures. Eventually, in 1975, the UN 'Convention concerning the Protection of World Cultural and Natural Heritage' came into force.

UNESCO define two types of site: cultural (which represent some aspect of human activity) and natural (which can be biological or geological). Mixed heritage sites contain both aspects: the volcanic archipelago of St Kilda is an example—it is home to endangered seabirds, so it is a site of importance in the natural world, but it also bears evidence of 2000 years of human occupation. Of the 1154 sites currently listed 897 are cultural, 218 are natural, and 39 are mixed.

But enough of the history of UNESCO. Let's cut to the important question: if you want to visit world heritage sites where should you fly to (pandemic restrictions permitting, and assuming you are comfortable with air travel—see Maps 22 and 23)? The five countries with the most sites are:

Italy	58
China	56
Germany	51
France	49
Spain	49

We might choose to assign two world heritage sites, the *Vatican City* and the *historic centre of Rome, the properties of the Holy See in that city enjoying extraterritorial rights and San Paulo Fuori le Mura*, to the city state of the Vatican rather than to Italy. If we do that then Italy and China host the same number of world heritage sites. But China, of course, is a much larger country: it has an area about 32 times larger than Italy. So if you want to hop from site to site then you're going to have to do a lot more travelling in China than in Italy. Representing the density of heritage sites for each country, however, is problematic. If one calculates the density of heritage sites per square kilometre, and one chooses to round down rather than round up (so 0.44 km^2, for example, becomes 0 km^2 rather than 1 km^2), then the calculation blows up: the density of world heritage sites in the Holy See is infinite! (It packs those two world heritage sites into an area of just 0.44 km^2.) Some other small states have a high density of heritage sites: Malta, for example, covers an area of just 316 km^2 but is home to the city of *Valletta*, the *megalithic temples of Malta*, and the *Hal Saflieni Hypogeum*. So rather than plot the density of sites this map just gives the locations.

You can spend a pleasant hour or two glancing through UNESCO's compilation of world heritage sites. If, for example, you ask yourself: 'what is the northernmost world heritage site?' then you treat yourself to some fascinating lessons in biodiversity and astronomy.

The site closest to the North Pole, it turns out, is one of the natural heritage sites: the *Natural System of Wrangel Island Reserve*. This consists of two islands (Wrangel and Herald) and the surrounding waters. Wrangel, because it escaped glaciation during the Quaternary Ice age, has an exceptionally high level of biodiversity for the Arctic: it is home to Earth's biggest population of Pacific walrus, has highest density of ancestral polar bear dens, and is a nesting ground for dozens of migratory bird species.

The northernmost cultural world heritage site is the *Struve Geodetic Arc*. The site actually consists of components at 34 separate locations. The most northerly is at Hammerfest in Norway but the Arc stretches down through ten countries. The site takes its name from the German–Russian astronomer Friedrich Georg Wilhelm von Struve who, in addition to carrying out careful observations of the sky, initiated a painstaking triangulation survey that established the precise size and shape of our planet. The survey, carried out between 1816 and 1855, used 265 main station points; the present-day *Struve Geodetic Arc* celebrates 34 of those original station points.

The most northerly site that gains its status for mixed cultural and natural reasons is the *Laponian Area*—a part of northern Sweden that is the world's largest unmodified nature area still cultured by natives. About 3000 Saami people (whom the English have historically referred to as Lapps) herd reindeer in this region.

The southernmost world heritage sites offer similar inspiration.

The site closest to the South Pole is one of the natural heritage sites: *Macquarie Island* is an uninhabited island discovered by Frederick Hasselborough in the southwestern Pacific Ocean in 1810. The island takes its place on the list because of its geological significance: it is the only place where rocks from the Earth's mantle, from 6 km below the ocean floor, are being actively exposed above sea-level.

The southernmost cultural world heritage site is the *Cueva de las Manos* ('Cave of Hands') in the Santa Cruz province of Argentina. Ancient artists stencilled hand prints on the cave walls by blowing paint through pipes made of animal bone onto outstretched hands (see Fig. 32). Some of those hands touched the walls 13,000 years ago.

Fig. 32 Over millennia, several waves of people have adorned the Cueva de las Manos with artwork. The artists painted the surrounding cliff faces too. The artwork typically depicts the guanaco (a llama-like creature that people hunted for food); the human body; and the human hand. (Pablo Giminez, CC BY-SA 2.0)

The most southerly mixed site is the *Tasmanian Wilderness*—a vast area that includes numerous important National Parks as well as archaeological plots that provide evidence of Aboriginal existence in these parts as long as 35,000 years ago. See Fig. 33.

UNESCO classify 52 sites as endangered. Many threats are predictable: war, the climate emergency, poaching. Urban development is another threat. Indeed, the UK was warned that developments on Liverpool's waterfront would cause the city to lose its heritage status; the developments went ahead anyway and in 2021 *Liverpool—Maritime Mercantile City* lost its world heritage status. (So far, the only other sites to have been delisted are the *Arabian Oryx Sanctuary* in Oman and the *Dresden Elbe Valley* in Germany.) Sites not even on the danger list can find themselves threatened. The *Tasmanian Wilderness*, for example, was threatened in 2014. The Australian government proposed the site be delisted so trees could be logged. No nation has ever had a site delisted for economic reasons and, after the World Heritage Committee rejected the proposal, the government backed down. The fact the proposal was made seriously, however, demonstrates that these sites are vulnerable to political manoeuvering. And of course some sites can be the victim of accident. In April 2019, for example, the Cathedral of Notre Dame—the medieval centrepiece of the World Heritage Site *Paris, Banks of the Seine*—succumbed to fire. At the time of writing it seems as if the cathedral will be rebuilt to its former glory, but the fire demonstrated clearly how quickly these treasures can be damaged.

Some of our planet's most evocative placenames—Damascus, Jerusalem, Timbuktu—are endangered. If they vanish under our stewardship would future generations forgive us?

Fig. 33 Cradle Mountain in the Tasmanian Wilderness. The wilderness region, which covers more than 15,000 km², underwent severe glaciation in the past. The last period of glaciation ended about 10,000 years ago. World heritage status is based on meeting one or more of ten criteria; the Tasmanian Wilderness meets seven of them, which along with Mount Tai in China is more than any other heritage site on Earth. There are few inhabitants now, but humans have lived in this region for tens of thousands of years. (Andrew Goddard, CC BY-SA 4.0)

34   Roman Amphitheatres (Map 34)

I have been involved, in a small way, in consultations about my university's new flagship building. Responsibility for its design lies with a talented team of architects, but academics have to inform the architects of how the building should function. The structure ought to have a useable lifespan of 70 years, and it is a pleasing thought that I will have had a (tiny) part to play in influencing a city's skyline long after I am gone. The cynic in me, though, wonders whether the new building really will be around seven decades from now. UK universities embarked on a post-millennial construction spree, and several developments that opened with a 'wow factor' are now already looking dated. I have even seen the demolition of buildings I watched going up. Buildings should last longer than people, but that's not always the case.

Such pessimistic thoughts are swept away when one visits Rome. In particular, the Colosseum has survived almost 2000 years of stone robbers, earthquakes, and thieves. It stands in disrepair, but little imagination is needed to conjure up a scene of 50,000 spectators cheering on their favourite gladiators. Modern football stadia, of which more in Map 40, are unlikely to last a fraction of the time the Colosseum has stood. (As a child I watched my own football team, Middlesbrough, at its Ayresome Park ground; bulldozers pulled the stadium down 94 years after its construction. The Colosseum still stands 1949 years after emperor Vespasian diverted funds from the spoils of the Jewish Temple for his Flavian amphitheatre.)

About 240 Roman amphitheatres have been identified, some of which remain impressive (and are world heritage sites; see Map 33). The Arles amphitheatre, which went up about 20 years after the Colosseum, and the Pula amphitheatre (see Fig. 34), which is slightly older than the Colosseum, are still used as entertainment venues. The ravages of weather, people, and earthquakes, however, have wrecked most amphitheatres. The most northerly example, Trimontium, close to the modern Scottish town of Melrose, is now visible as just a grassy hollow. I suspect in 2000 years my university's newest building will leave even less trace.

34   Roman amphitheatres

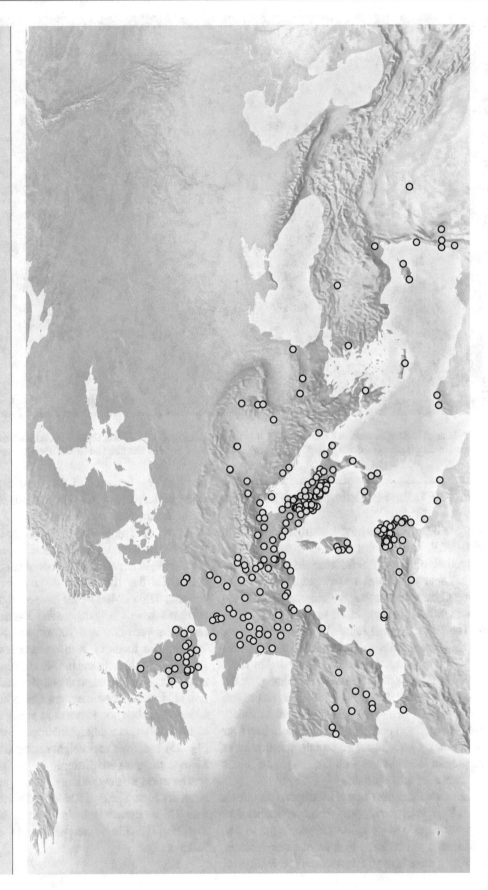

Map 34

Fig. 34 The Pula amphitheatre, in modern Croatia, was built in the first century AD. In Roman times it hosted gladiatorial games watched by 20,000 spectators. In the middle ages, cattle grazed here. Now, it is a venue for concerts and film festivals. (Jeroen Komen, CC BY-SA 2.0)

35   The 100 Tallest Buildings (Map 35)

The Great Pyramid of Giza, whose construction began in around 2580 BCE, and which served as a tomb for the Ancient Egyptian pharaoh Khufu, reached a height of 146.5 m. I am sure later kings and emperors dreamed of surpassing the Great Pyramid, but the construction of anything taller was difficult because of the technology available to masons. The Great Pyramid remained the world's tallest building for almost four millennia, unsurpassed until workers completed the 160 m-tall spire of Lincoln Cathedral in 1311. Lincoln Cathedral itself held the 'tallest ever building' record for more than 500 years. Since about 1890, the record has changed with greater frequency. Technological advances have allowed architects to pander the ego of those in power and design ever-taller buildings. (To be fair, practicality as well as ego plays a role: in cities with limited ground area a skyscraper offers a lot of real estate.)

A building much taller than Lincoln Cathedral requires a steel framework on which curtain walls can hang, rather than load-bearing walls made of stone. And if people are to live and work in such a tall building then an elevator becomes much more than a mere luxury. The necessary technological developments occurred in the nineteenth century so that is when skyscrapers, as we now understand them, began to be built—for the most part in America.

As of 2021, more than 300 skyscrapers taller than 300 m are either built, under construction, or in the planning phase. As you might expect, the number of tall buildings drops with height. Of the 100 tallest skyscrapers 59 have a height between 338 and 400 m; 29 have a height between 400 and 500 m; 8 have a height between 500 and 600 m; 3 have a height between 600 and 700 m; and only 1 is taller than 700 m. The focus of such construction has shifted away from America to Asia and the Middle East, with most activity taking place in China and the oil-rich states. The transformation of skylines in southeastern China in particular is astonishing. Shenzhen, the fourth most populous city in China, has eight of the world's tallest buildings; nearby Hong Kong (see Fig. 35) has five; and neighbouring Guangzhou has four. More skyscrapers are planned.

The world's tallest building, though, at the time of writing, is not in China. The Burj Khalifa in Dubai, at a height of more than 828 m, dwarfs Dubai's 11 other supertall skyscrapers. The Burj Khalifa is so tall its tip can be seen from 95 km away.

35 The 100 tallest buildings

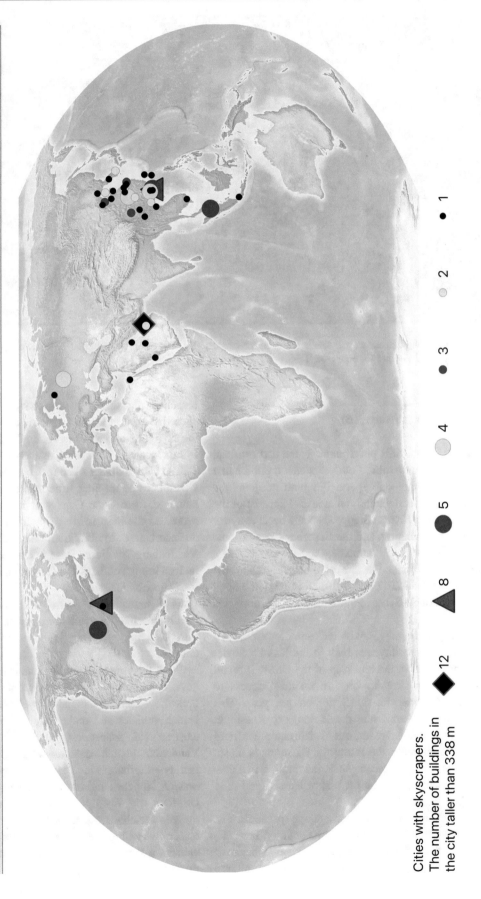

Cities with skyscrapers.
The number of buildings in
the city taller than 338 m

◆ 12
▲ 8
● 5
● 4
● 3
● 2
• 1

Map 35

Fig. 35 The Hong Kong harbour and skyline at night. In the two decades between 1990 and 2010, Hong Kong erected 5 of the 100 tallest skyscrapers in the world. (Benh Lieu Song, CC BY-SA 4.0)

36   Spaceports (Map 36)

In October 1957 the USSR put Sputnik 1, the first artificial satellite, into a low Earth orbit. Four years later it put the first human—Yuri Gagarin—into space. For both missions the point of launch was in Baikonur, an area 200 km east of the Aral Sea in the harsh Kazakh Steppe. The site became a new type of facility, a spaceport, and the Baikonur Cosmodrome remains the world's largest launching site for space. Since 1957 Baikonur has been the site of almost 1400 orbital launches and, between 2011 and 2019, all astronauts bound for the International Space Station launched from this spaceport.

The area surrounding the Baikonur Cosmodrome enjoys a strange history. The facility was built in the Kazakh Soviet Socialist Republic; after the break up of the USSR it found itself in Kazakhstan. Russia, keen to maintain access to space, rented Baikonur for $115 million per year. This 6500 km² patch of Kazakhstan thus belongs to Russia, at least until 2050. But pitted roads, out-of-date equipment, and stagnating facilities mean the future for Baikonur is bleak. Roscosmos, the Russian space agency, is investing in a new spaceport at Vostochny, in the far east of Russia (and fully within Russia's borders).

Russia also claims the world's most active spaceport. The Plesetsk Cosmodrome, in the north of the country, has been the site of more than 1500 orbital launches. Plesetsk began as a military facility and it remains better suited for military

launches: reaching a geostationary orbit, the commercially attractive orbit in which a satellite appears at a fixed point in the sky relative to an observer on the ground, is uneconomic from a location as far north as Plesetsk.

The United States, of course, has spaceports. In 1962 John Glenn became the first American to orbit the Earth, and he launched from Cape Canaveral in Florida. The Cape also served as the base for the Apollo program and now offers an alternative to Baikonur for reaching the ISS. The Vandenberg Air Force Base is another busy spaceport, but its location in southern California limits the types of satellite it can economically deploy.

The Guiana Space Centre, which lies near the equator, is ideally located for launching geostationary satellites. The European Space Agency launches its Ariane rockets from Guiana.

What this map fails to illustrate is how private companies have begun to explore new options for reaching space. SpaceX, a company founded by Elon Musk, manufactures launch vehicles and rocket engines. SpaceX operates from existing US spaceports, such as Cape Canaveral and Vandenberg, but the company is also building its own launch facility—Starbase—in Texas. A key difference between SpaceX launches and those early launches by the USA and USSR is the way that much of a modern vehicle can be reused. Watching the launch of a SpaceX rocket inspires respect; watching the controlled return of a reusable launch vehicle, a landing that can either be at a spaceport or on a floating pad in the ocean, inspires awe.

36 Spaceports

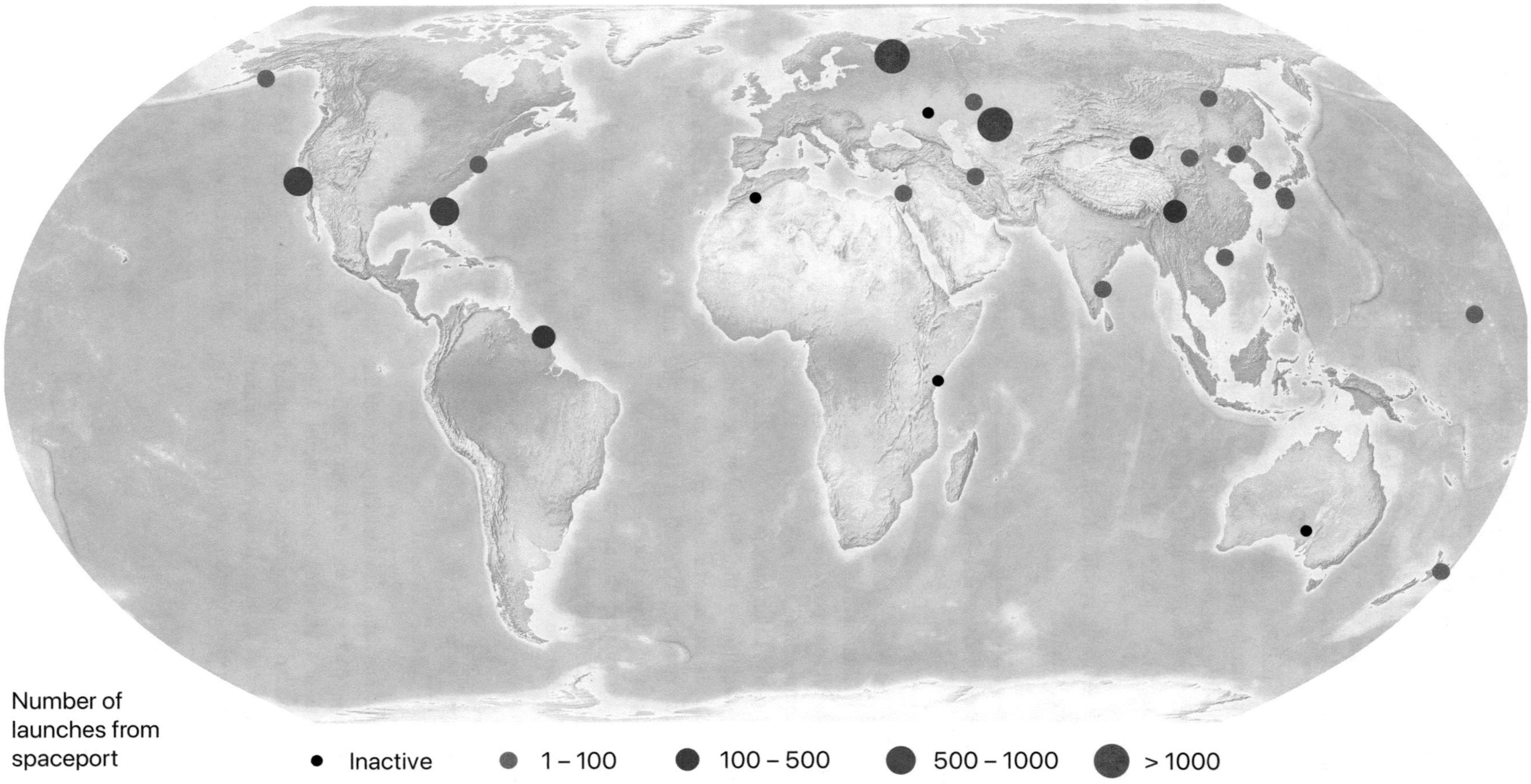

Number of
launches from
spaceport

Inactive ●

1 – 100 ●

100 – 500 ●

500 – 1000 ●

>1000 ●

Map 36

Another billionaire, Jeff Bezos, is also reaching for space. His Blue Origin company operates a suborbital launch site in Texas (the site from which William Shatner, the actor who played *Star Trek*'s Captain Kirk, launched in 2021 to reach the edge of space). The site is not a fully fledged spaceport, so is not shown on the map, but Blue Origin plans to develop a new orbital launch facility at Cape Canaveral. In years to come we can surely expect to see more spaceports built by private companies.

China, meanwhile, is building a new facility—the China Oriental Spaceport—in Shandong province. Alongside this it is building a 162.5-m-long, 40-m wide ship that will be capable of launching rockets into space. This mobile launch platform will allow China to launch from wherever happens to be most convenient.

Spaceflight languished after Apollo. Right now, space appears closer than ever before.

---

37   The 500 Most Powerful Supercomputers (Map 37)

From 1986 to 1992, Hans-Werner Meuer, a computer scientist at the University of Mannheim, collated information on the world's most powerful supercomputers and presented his results at an annual conference. In 1993, along with Jack Dongarra, Horst Simon, and Erich Strohmaier, he initiated a formal approach to establishing the 500 most powerful machines. Twice each year an international group of computer science experts compile TOP500—a list of the 500 fastest, commercially available, general-purpose supercomputer systems based on performance on a benchmark test. Every 6 months the list is updated, and every 6 months the average speed of the computers increases.

> As long as we can make them smaller, we can make them faster.—Seymour Cray (designer of the first supercomputers)

In June 2019, for the first time, all supercomputers on the list delivered a petaflop or more on the benchmark statistic. The petaflop is a unit of computing speed, useful when discussing high-performance computing. A computer operating at 1 petaflop can deliver 1000 million million operations using floating-point arithmetic *every second*. To

get an idea of how rapidly the field progresses, the most powerful supercomputer on the first TOP500 list in June 1993 was rated at just under 60 gigaflop—more than 7 million times slower than Fugaku, a Japanese supercomputer that is, as of November 2021, the world's fastest.

The geographical spread of supercomputers is perhaps as one would expect. As of November 2021, China has the largest number (173), followed by the USA (149), with third place belonging to Japan (32) or to Europe as a collective. If one looks instead at total supercomputing performance, which is perhaps a more useful metric, then the picture changes slightly: the US has 32.5% of the performance share followed by Japan (20.7%) then China (17.5%).

This map is important because whoever possesses supercomputing power is in possession of a powerful tool. Here is a list, off the top of my head, of some of the uses to which that power can be put. Biochemists use supercomputers to help determine the structure of compounds; chemists use them to compute the properties of biological macromolecules; climatologists use them for climate research; cryptanalysts use them for code breaking; engineers use them for understanding air flow over structures; geologists use them for fossil fuel exploration; material scientists use them to model the behaviour of airplane components; meteorologists use them for weather forecasting; physicists use them for probing quantum mechanics; weapons scientists use them to model nuclear detonations. Japanese scientists will use Fugaku, that current world-leading 442 petaflop supercomputer, to identify potential treatments for Covid-19 and to map escape routes from tsunamis.

Finally, researchers are using petaflop supercomputers to understand how to build exaflop supercomputers—machines with the same sort of processing power of the human brain. Fugaku has almost reached the exascale. For applications involving machine learning and artificial intelligence, numbers typically do not need to be stored with the same sort of precision as is needed in many other applications. When Fugaku uses this reduced precision for storing numbers it can reach a peak performance of just over 1000 petaflop—in other words, 1 exaflop. The November 2022 edition of the TOP500 list saw the Frontier computer (at Oak Ridge National Laboratory, USA) knock Fukagu off top spot. Frontier can reach 1.1 exaflop.

Exascale computing is here.

37 The 500 most powerful supercomputers

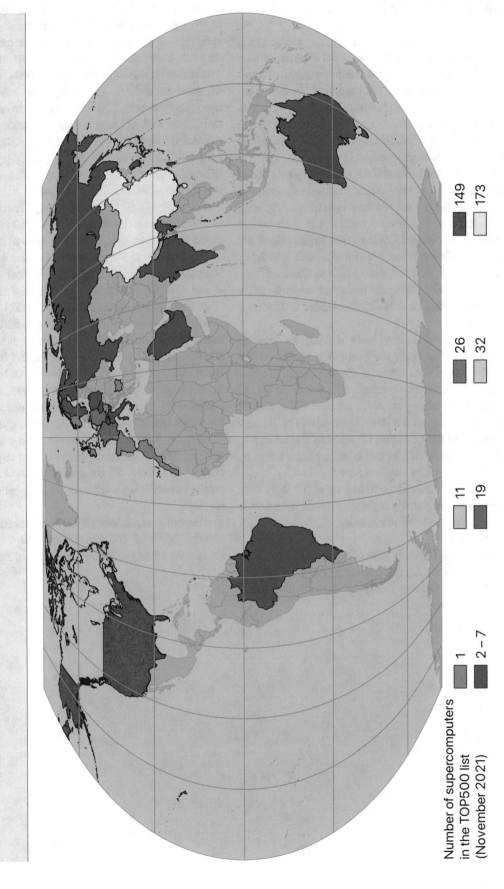

Number of supercomputers
in the TOP500 list
(November 2021)

1
2 – 7
11
19
26
32
149
173

Map 37

38   Mains Frequency (Map 38)

Every traveller to a foreign land inevitably encounters a problem: how to fit the plug of an electrical appliance into a wall socket. The world uses a bewildering variety of plug-and-socket combinations. In the USA one finds sockets compatible with a type A plug (two flat parallel prongs) or type B (two flat blades with a rounded earth pin). In the UK we typically use a socket compatible with plug type G: three pins of rectangular cross section arranged in a triangle, with the topmost pin aligned vertically and the other two pins aligned horizontally. Countries in the Arabian peninsula also tend to use plug type G, as do some African countries along with Malaysia, Singapore, and several others. The alphabetical fact that there are A, B, and G plug types implies there are at least six types. The list in fact goes up to type O, which has three rounded pins, in an arrangement unique Thailand.

At least the traveller's situation is simpler when it comes to mains frequency (or line or utility frequency, as they refer to it in America).

All countries use alternating current (AC) in order to transmit electric power from the generating station to the customer. The world then splits nicely into two. In most countries the AC oscillates 50 times per second: 50 Hz, in other words. The remaining countries produce AC at 60 Hz. (The grid frequency is never a pure 50 Hz or 60 Hz. Rather, there is a slight variability around the nominal frequency—a fact that forensic scientists can use to authenticate audio or video recordings for use in criminal investigations.)

Why is there any difference in frequency at all? Why does the world not agree on a single frequency standard, and stick to it? The answer lies in the history of the electricity generating industry.

In the USA, during the early days of electricity generation, the utility companies did not use alternating current. Edison's pioneering General Electric Company instead distributed direct current (DC). The advantages of AC over DC began to win out, but a variety of frequencies were in use; the operators of steam engines and water turbines would use the most convenient frequency for their machines, and that would vary. It was only after Nikola Tesla devised his system of three-phase AC electricity (in other words, three alternating currents, 120° out of phase, to smooth out voltage variations) that AC won a decisive victory over DC. Tesla calculated that 60 Hz was the most effective frequency for his particular equipment, and this frequency gradually became a standard—but for no deep reason. The standardisation is merely a fossilised accident of history.

The choice of 60 Hz spread through most of the Americas, as can be seen on the map. The use of 60 Hz in Saudi Arabia seems strange, though, given that none of its neighbours use it. Presumably the explanation lies in that American oil companies developed the Saudi electrical power grid. They would naturally use equipment operating at 60 Hz.

In Europe, the German company AEG started what was almost a monopoly on electricity generation. The German engineers thought 50 Hz was a better metric standard than 60 Hz, and again this accident of history became fossilised. European countries of course had links to other countries around the world, links often forged through war or colonisation but nevertheless strong enough to determine the mains frequency. Most of the world uses 50 Hz.

In truth, there is little to recommend one frequency over another. The 60 Hz transformers can be slightly smaller and cheaper than their 50 Hz counterparts. On the other hand, power transmission over long lines is better at 50 Hz than 60 Hz.

The other consideration for the traveller is, of course, voltage. Most of the world uses 220–240 V. The USA, along with a few other countries, prefers 120 V.

38   Mains frequency

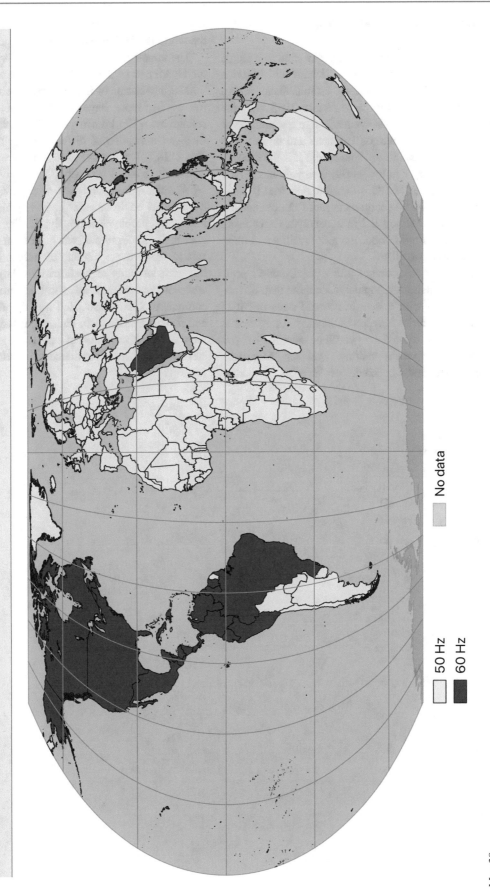

50 Hz

60 Hz

No data

Map 38

39   Sharing a McDonald's (Map 39)

The McDonald's fast food empire began in 1940 as a single drive-in eatery founded by brothers Richard and Maurice McDonald. The brothers grew that establishment into a chain of restaurants (see Fig. 36), and in 1955 sold the chain to Ray Kroc. Under Kroc's leadership the number of these fast food outlets rocketed. The facts and figures now associated with McDonald's are staggering: it is the second-largest private employer in the world; the chain has served something of the order of 385 *billion* burgers; and it has a presence in well over 100 countries (a fact I'm occasionally grateful for because, although I'm squeamish about using their touchscreens to order food, the toilet facilities in these outlets are generally excellent).

Not every country encourages a large McDonald's presence. In Iraq, for example, a population of more than 38 million must make do with a lone McDonald's restaurant in Baghdad. Some countries have no McDonald's restaurant at all. Most of Africa lacks an outlet; you'll be out of luck if you want a McFlurry when visiting Iran or Mongolia; McDonald's restaurants in countries such as Barbados, Bermuda, and Bolivia have closed down. But in much of the world the McDonald's logo is ubiquitous.

The largest ratio of restaurants to people can be found in the Dutch part of the Caribbean island of Saint Marten. The 40,120 inhabitants of Sint Maarten, a constituent country of the Netherlands, have a choice of three McDonald's restaurants. (The French part of the island is McDonald's-free—an indication of differing French and Dutch tastes?) The USA, of course, is well served with on average one McDonald's outlet for every 23,130 inhabitants.

When any company enjoys a global reach we should examine its environmental impact. McDonald's takes environmental problems seriously: it ditched the polystyrene clamshell, for example, when customers voiced concerns. But to make those billions of burgers the company must buy vast amounts of beef. In turn, large amounts of land and water must be devoted to beef production—with concomitant problems associated with deforestation (see Map 8) and methane production. Unless our taste for meat changes, there is a limit to what McDonald's or any similar company can do to limit its environmental footprint.

39   Sharing a McDonald's

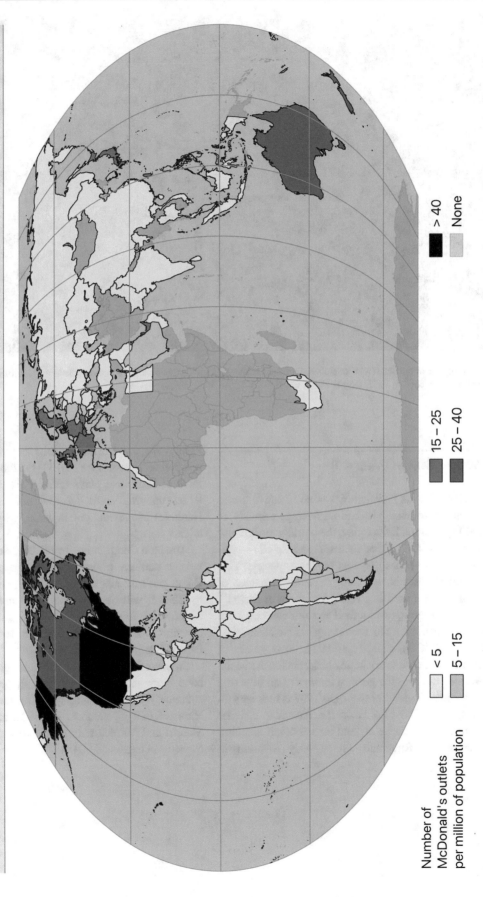

Number of
McDonald's outlets
per million of population

- < 5
- 5 – 15
- 15 – 25
- 25 – 40
- > 40
- None

Map 39

Fig. 36 A photograph of the oldest McDonald's restaurant still in operation. The restaurant is located in Downey, California, and was the third to be built. The restaurant opened in 1953. (Bryan Hong, CC BY-SA 2.5)

40 Soccer Stadia (Map 40)

The Colosseum, over its five centuries or so of active use as an amphitheatre (see Map 34), had an average capacity of about 65,000 spectators. Even today, the world's most popular spectator sport—football, or soccer, (see Map 61)—typically takes place in venues that seat far fewer people. This map shows all stadia with a seating capacity in excess of 60,000 that have been used as a venue for games of football.

At the time of writing, two football grounds can seat more than 100,000 spectators. As its name suggests the Melbourne Cricket Ground (MCG), capacity 100,024, hosts games of cricket. It is better known to cricketing historians as the venue for what Wisden ranks as the greatest innings of all time (the Don's 270 against England in the 1936/1937 Ashes series). Nevertheless, the MCG also hosts the Australia national football team and so can be classed as a football stadium. Even larger is the Rungrado 1st of May Stadium in Pyongyang, a venue mainly used for gymnastic performances and political events (see Fig. 37) but the North Korea national football team also plays there. This was once the world's largest sporting venue but now that accolade belongs to the Narendra Modi Stadium in India, which can seat 132,000 cricket fans.

The Nou Camp, home to FC Barcelona since 1957, is the largest stadium whose primary use is football. The Nou Camp can seat just shy of 100,000; an upgrade means that by 2024 it will overtake the MCG in terms of capacity. In addition to football, the stadium has been used for pop concerts, operas, and even church services (in 1982, John Paul II celebrated mass there for over 121,000 people).

An arc of large stadia snakes through Europe. Some cities have more than one large football ground. London alone contains Wembley (capacity 90,000) and grounds for West Ham (66,000), Tottenham (62,303), and Arsenal (60,338). Madrid and Munich both have two football grounds with a capacity in excess of 60,000.

40 Soccer stadia

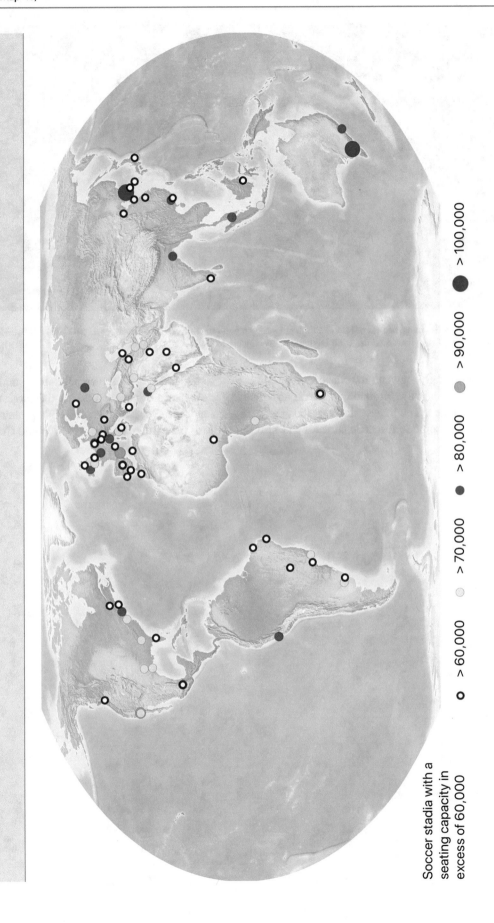

Soccer stadia with a
seating capacity in
excess of 60,000

> 60,000   > 70,000   > 80,000   > 90,000   > 100,000

Map 40

Fig. 37 The interior of the Rungrado 1st of May Stadium in Pyongyang, North Korea. The venue opened on 1 May 1989, and was built in part as a response to South Korea being awarded the 1988 Olympic Games. (Christophe95, CC BY-SA 4.0)

Perhaps surprisingly, given the historic success of Brazilian football, only five grounds in Brazil seat more than 60,000 spectators. The largest stadium, the Maracanã, capacity 78,838, is home to the national team and also the clubs Flamengo, Fluminese, Botafogo, and Vasco da Gama. The Maracanã played host to the largest official football attendance ever recorded: 199,854 spectators stood to watch Brazil play Uruguay in the 1950 World Cup final.

None of the structures discussed in the previous chapter appeared fully formed in the world. Someone first had to imagine them. The world of the mind is important not only because our thoughts help create our personal reality, but because, almost uniquely among living creatures, humans can transform imagination into substance. Humans have been astonishingly successful at implementing those transformations. Indeed, perhaps humanity's eventual undoing will be the conviction among certain politicians that those transformations are *always* possible, can *always* take place. The conviction that the external world must accord with what goes on inside their heads.

In this chapter we consider maps that illustrate some aspects of science. As much as any other community of people, scientists work with thoughts, ideas, and beliefs—but scientists test their views against reality. Science is a structured attempt to describe the world as it *is*, not as we would like it to be.

We consider maps that illustrate various aspects of education.

And we consider maps that illustrate religious belief and happiness. Whether there is a correlation between the two I leave the reader to decide.

41   The Most Complex Machine Ever Built (Map 41)

CERN—the European Organisation for Nuclear Research—was established in 1954 to facilitate world-class research into fundamental physics. The Geneva-based organisation asks deep questions, such as: what is the universe made of? And it attempts to answer them by allowing some of the world's brightest minds to design, implement, and analyse experiments using the world's largest particle physics laboratory. Although CERN itself is a European project—of the 23 full member states only Israel is outside Europe—a glance

at this map of just one experiment at CERN shows scientific cooperation operates here on a global scale.

CERN runs a number of machines, but the one surely everyone has heard of is a particle accelerator called the Large Hadron Collider (LHC).

The LHC, which has been described as the most complex machine ever built, has a descriptive name. The machine is *large*. (The circular tunnel, which contains a labyrinthine collection of superconducting electromagnets used to accelerate and shepherd electrically charged particles, has a circumference of 27 km. The tunnel lies 100 m underground, and runs beneath the border of France and Switzerland, between the Jura mountains and Geneva international airport. The bulk of its length is underneath French soil. See Fig. 38.) It accelerates *hadrons*. (A hadron is a type of subatomic particle made up of smaller particles called quarks. The two commonest hadrons are the proton and the neutron. The LHC accelerates protons to energies much higher than those achieved by any previous accelerator.) And it is a *collider* because it smashes together—or collides—those hadrons.

Why do physicists want to crash protons together? Well, although smashing particles into each other may seem destructive, it is currently our best way of answering the question 'what is the universe made of at the fundamental level?'

What is a truly elementary particle? Well, the natural world consists of just over 90 different substances, or elements (see Map 9). The elements, however, are not elementary. Elements consist of atoms; an element is merely a substance that consists of just one type of atom. But an atom has a complex internal structure. Atoms consist of a nucleus and shells of orbiting electrons, so atoms are not elementary either. Neither is the atomic nucleus: it too has a structure—it contains protons and neutrons. (In a neutral atom, the number of orbiting electrons is equal to the number of protons in the nucleus; there are varying numbers of neutrons. Atoms of different elements contain different numbers of protons and electrons. The hydrogen atom, for example, has 1 proton in

S. Webb, *Around the World in 80 Ways*, https://doi.org/10.1007/978-3-031-02440-5_6

41 The most complex machine ever built

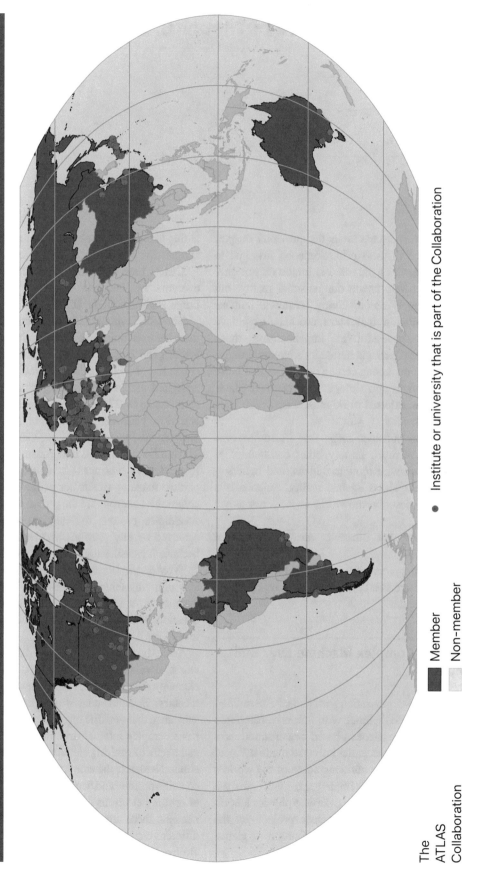

The
ATLAS
Collaboration

● Institute or university that is part of the Collaboration

Member

Non-member

Map 41

Fig. 38 An aerial view of CERN, which shows the Alps with Mont Blanc in the background and Lake Geneva in the midground. Superimposed on the photograph is a ring showing the location of the 27-km-long LHC tunnel. The position of four experiments—CMS, LHCb, ATLAS, and ALICE—is highlighted. (Maximilien Brice (CERN), CC BY 4.0)

the nucleus and 1 orbiting electron; the helium atom has 2 protons in the nucleus and 2 orbiting electrons; the lithium atom has 3 protons and 3 electrons; and so on up to uranium, which has 92 protons and 92 electrons.) The electron, as far as we know, has no internal structure; it *is* an elementary particle. But protons and neutrons are not elementary: they are each made up of three quarks. The quarks, as far as we know, *are* elementary: as with the electron, they have no internal structure.

Nature thus seems to possess a 'Russian doll' aspect, and physicists learn about it by colliding particles and seeing whether, and how, they break up. The higher the energy of collision, the smaller the scale at which physicists are able to probe the structure of matter.

At the LHC, collision energies are higher than anywhere else on Earth. (Well, they are higher than at any other *laboratory*. Every day, cosmic rays strike Earth's atmosphere

with energies that dwarf anything our technologies can achieve.) Protons are made to circulate in two separate beams travelling in opposite directions, and the magnets bend, focus, and squeeze the beams. At four points in the ring the proton beams are made to collide. Ensuring those collisions occur is a feat in itself: the beams are so narrow, this is like firing two needles 10 km apart and having them hit each other at the halfway point. Four different detectors— ALICE, ATLAS, CMS, and LHCb—study the outcome of the collisions.

Each detector occupies a vast underground cavern. The two largest experiments, ATLAS (A Toroidal LHC Apparatus) and CMS, are general-purpose discovery machines; they were designed independently and are operated by different teams. This allows physicists to confirm the authenticity of a discovery. If ATLAS *and* CMS measure the same effect then we can be confident the effect is real rather than a statistical

Fig. 39 With a length of 46 m and a height and width of 25 m the ATLAS detector at the LHC is the largest particle detector ever built. A single photograph such as this can only hint at the complexity and sophistication of the experiment. (Simon Waldherr, CC BY 4.0)

fluke. ALICE and LHCb are smaller detectors, used to study specific phenomena. Let's focus here on ATLAS.

The intricacy and complexity of the ATLAS detector (see Fig. 39) is necessary because of its function: the collisions inside it create myriads of particles, whose type must be identified and whose momentum and energy must be measured. The experiment generates *vast* amounts of data. Almost as complex as the detector is the collaborative effort required to operate the experiment and produce good science. The ATLAS Collaboration is one of the world's largest scientific efforts. At the time of writing, the Collaboration contains about 3000 scientists from 183 institutions representing 38 countries. About 1200 doctoral students help with the development of the detector, data collection, and data analysis. And none of this could happen without the support of engineers, technicians, and administrative staff. In a world that holds so much division, I find comforting that an international partnership can work so successfully.

And the ATLAS Collaboration *has* been successful. Its outstanding achievement has been to find evidence for the Higgs boson, the final missing piece of the so-called Standard Model of Particle Physics. Back in the early 1960s a number

of physicists, including Peter Higgs, postulated the existence of a field that permeates the entire universe. They argued that different elementary particles would interact with this field in different ways. The elementary particle we call the photon, for example, would be unaffected by the field; the elementary particle we call the W, on the other hand, *would* be affected. The idea was that both the photon and the W are inherently massless. The photon, being aloof and not interacting with this postulated Higgs field, remains massless; the W, because it interacts with the Higgs field, becomes massive. This mass-generation mechanism permitted physicists to write down equations that applied to both the photon and the W, and at the same time explain why we observe the photon to be massless and the W to be massive. Higgs showed that if there were indeed a universal field, whose interactions with elementary particles could generate the mass of those particles, then under certain circumstances the field could be "excited" and a particle associated with the field would be generated. This field excitation, this particle, was named the Higgs boson—and on 4 July 2012 the ATLAS and CMS collaborations announced they had found evidence for it. Higgs and two of his colleagues, Robert Brout and

François Englert, were awarded the Nobel prize in 2013, about four decades after they made their initial suggestions.

Physicists at ATLAS are now on the search for more exotic entities. Some researchers hope that by smashing particles together at even higher energies than before they might find exotic phenomena: extra dimensions, perhaps, or microscopic black holes.

42 Scientific Productivity (Map 42)

The weekly journal *Nature* is perhaps the pre-eminent publication for the dissemination of scientific research. In recent years the journal has advanced the study *of* science, and the culture of science, through its development of the *Nature Index* database. The *Index* compilers comb through papers appearing in an independently selected group of 82 high-quality life- and physical-sciences journals and collate data on author affiliation. Those journals constitute less than 5% of the number of journals published *in* the natural sciences, but they deliver almost a third of the total number of citations *to* natural science journals. The *Index* thus provides a measure of the quality of science taking place at the institutional and country level. (It does not indicate the quality of scientists produced by different educational systems. For example, a paper might have four authors with MIT as their affiliation, four with Oxford as their affiliation, and two with Heidelberg as their affiliation. That would count as a score of 4 for the USA, 4 for the UK, and 2 for Germany. But *none* of those authors need necessarily be American, British, or German nationals. Science is an international endeavour; the best scientists tend to go where the best science is taking place.) So which countries are doing well in the *Index*?

It comes as no surprise to learn the USA publishes the largest number of high-quality papers. China is a distant second, Germany third, the UK fourth. (A competitor to the *Index*, the Scimago project, aims for breadth of coverage rather than quality. It draws data from more than 34,000 titles, and so indicates how *much* science is taking place within a country. Under this scheme the USA remains first,

with China a much closer second, and the positions of the UK and Germany are reversed.) A more interesting picture emerges when one considers the number of high-quality papers a country publishes when account is made population size.

Just as overall numbers favour large countries, per capita numbers favour small states. An unexpected result, at least to me, is that the Vatican has the biggest per capita output of high-quality scientific papers. The Holy See is no hotbed of scientific activity, but it does host the Pontifical Academy of Sciences (which counts some great scientists among its ordinary members) and the Vatican Observatory. In 2019, the year under review here, the Vatican Observatory appeared as the affiliation on six high-quality papers (on the subjects of gravitational waves and globular clusters, if you want to know). The Vatican's output per person was more than ten times that of Monaco, the second most productive territory. Members of the Scientific Centre of Monaco were authors on 19 high-quality papers; again, the principality's small population gives it a large papers-to-people ratio.

If we exclude the Vatican and Monaco, Switzerland heads the list. Switzerland plays host not only to CERN (Map 41) but also to ETH Zurich, one of the world's leading universities. Einstein was a student and professor there, and numerous other Nobel prize level graduates and professors have been affiliated with it. Several other northern European countries are productive. In addition to obvious big hitters such as the UK and Germany, Greenland appears high on the list. (Greenland's small population is well placed to undertake research in natural resources.) Outside Europe, high-ranking countries include Singapore, Israel, and Australia.

Few African countries host institutions to which the authors of high-quality scientific papers are affiliated. Quite how this situation can be improved is difficult to say, and yet Africa surely needs to build its science base. Africa needs scientists to help tackle the increasing challenges of drought, famine, and species loss. But a talented African scientist is likely to flee to a country that already has the infrastructure, laboratories, and libraries needed to support science. (Or, which is even more likely, that scientist will be lost to science completely.)

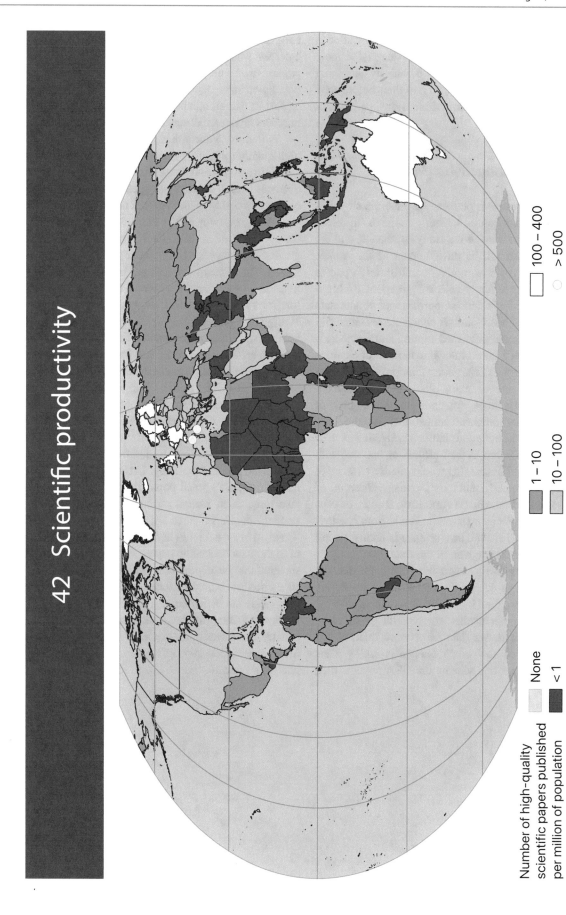

42 Scientific productivity

Number of high-quality
scientific papers published
per million of population

None

<1

1 – 10

10 – 100

100 – 400

> 500

Map 42

43   Intellectual Property (Map 43)

Societies began exploring the notion of intellectual property (IP) rights more than 2500 years ago. The leaders of the ancient city of Sybaris, for example, allowed the discovers of 'any new refinement in luxury' to enjoy the profits accruing from that discovery for a period of 1 year. (The law appears to have been aimed at protecting the recipes of bakers or chefs. The people of Sybaris certainly enjoyed their feasts, and even today we use the word 'sybarite' in that hedonistic sense.) The idea of a patent was, and remains, a simple one: it gave its owner the right not to make or sell their invention themselves but to *exclude others* from making or selling that invention for a specified time. The intention was to encourage people to invent, experiment, discover—and to feel confident they could share their findings with society, thus increasing the public good while still profiting from their work. After Ancient Greece, rulers at various times and in various countries granted 'letters of patent' to inventors. And in 1474, in Venice, a codified system of patent rights came into being. Venice established the basic principles of modern patent law.

The Swiss-based World Intellectual Property Organization (WIPO), a self-funding agency of the UN, is a global forum for intellectual property services, policy, information, and cooperation. By helping to protect intellectual property it aims to foster an environment in which innovation can thrive. (The phrase intellectual property refers to *any* creation of the mind—not just inventions, but also literary and artistic works, designs and symbols, and commercial names and images. Patents are just one way in which IP is protected in law. Copyright and trademarks are other examples.) WIPO also collects statistics on the granting of patents at the national level, which I show here. Given the distortions to various economic activities caused by the Covid-19 pandemic, I have chosen to look at the situation regarding patents just before the coronavirus struck.

If we exclude San Marino, whose 33,400 inhabitants were granted 686 patents by their national office in 2018, the per capita distribution of patents is what one might expect. A hotbed of patented invention lies in the East, with South Korea being the most inventive nation (2305 patents granted per 100,000 of population), followed by Japan (1537), and then Hong Kong (1295).

Fourth on this list comes the USA, with 941 patents granted per 100,000 of population in 2018. The large size of its population, however, means the USA granted a large number of patents in total: 307,759. Only China granted more patents that year. At the other end of the list, many countries—especially those in Africa—granted no patents at all via a national office.

Does it matter, this disparity in patent activity between Hong Kong and Haiti, say, or South Korea and South Sudan? Well, for an invention to be granted a patent it must not only be novel and non-obvious, it must also also serve a useful purpose. So a patent might be awarded for an innovative business method; new computer software or hardware; an ingenious machine; a different type of plant; and so on. The award of the patent enables the inventor, whether an individual or a company, the chance to profit from the invention. But the invention itself might have the potential to improve people's lives more broadly and lead to increased economic activity. The World Bank classifies countries as high income, upper-middle income, lower-middle income, or low income. In 2018, the high income and upper-middle income countries were granted 97.6% of all patents. If this is any indicator of future economic performance, the gap between rich and poor countries will remain or widen.

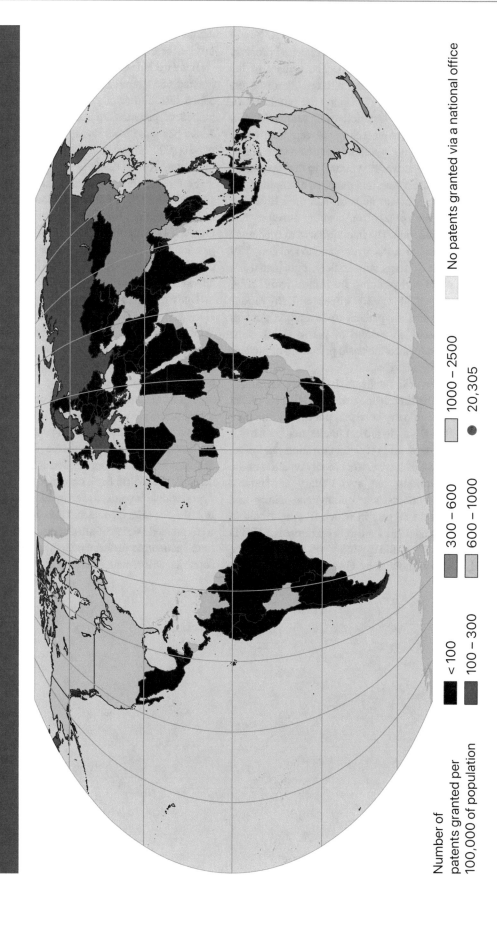

43 Intellectual property

Number of
patents granted per
100,000 of population

< 100

100 – 300

300 – 600

600 – 1000

1000 – 2500

20,305

No patents granted via a national office

Map 43

44   Views on the Safety of Vaccines (Map 44)

Measles is a highly contagious disease: a patient typically infects 12–18 other people. When I was at school, if one kid succumbed to measles then every kid soon had it. I caught it at age 6 and, as happens in most cases, I quickly recovered. But some patients suffer complications. For every thousand children who get measles one will develop an inflammation of the brain. In 1962 Olivia, eldest daughter of the author Roald Dahl, died from measles encephalitis.

The measles vaccine was introduced in 1963. It is effective and safe. Even better, if 90–95% of a population is vaccinated then 'herd immunity' comes into play—the virus no longer spreads easily, so even the unvaccinated (those who are perhaps too young, or who have a suppressed immune system) are unlikely to catch the disease. For a less contagious disease such as polio, herd immunity requires only 80–85% of people to be immunised. So why, when we have cheap, effective, and safe vaccines, are immunisation rates starting to fall?

The Wellcome Trust, a wealthy charitable foundation, commissioned Gallup to administer the world's largest study into how people feel about science and their attitudes to major health challenges. In 2019, they published the results—an analysis of responses from over 140,000 people from more than 140 countries. The most surprising element for me was the finding that one in three people in France *disagreed* with the statement 'Vaccines are safe'. In some other European countries—Austria, Belgium, Iceland, Switzerland—more than one in five people disagreed with the statement. This scepticism might not necessarily translate into intended action, but it raised the concern that vaccination rates would drop, herd immunity would fail, and children would once again start dying of a common childhood disease such as measles—as is happening in other places around the world.

The Covid-19 pandemic put those attitudes to the test. At the beginning of 2020, as the magnitude of the health crisis facing the world became apparent, the accepted wisdom was that a vaccine for Covid-19 was years away. An almost miraculous application of science meant that, by the end of 2020, rich countries had several highly effective vaccines from which they could choose. It is one thing for a person to register scepticism in a survey; would that scepticism translate into a refusal to protect oneself, one's family, and society in general?

At the time of writing, despite the high level of scepticism expressed in that Wellcome Trust survey, a large majority of the French population has been fully vaccinated against coronavirus. The vaccination rate in France is not quite as high as in Spain or Portugal, for example, countries that—as can be seen on the map—had lower rates of vaccine scepticism. But it is higher than in Austria, German, or Switzerland, countries that did not express quite the same level of scepticism in the survey. For whatever reason, it seems many people in German-speaking countries are vaccine hesitant. In the USA, an initially effective vaccine roll-out stalled: a mix of conspiracy theories and divisive politics seems to be responsible. In the UK and several other countries, a hard core 'anti-vax' sentiment exists.

Tragically, Covid-19 is becoming a disease mainly of the unvaccinated.

How can we change attitudes to vaccination, an intervention doctors have used effectively for over 250 years? A natural response is to say we should educate people. But perhaps this deficit model of communication, the idea people behave differently when they have more information, is ineffective. Or at least insufficient. Maybe we need better storytellers. Perhaps we need to republish the anguished essay Roald Dahl wrote after the death of his daughter.

The map illustrates one more stark result. People in some countries agree strongly that vaccines are safe. Yet many of those same countries are poor. Their citizens have not had the chance to enjoy the same level of protection offered to the citizens of richer countries.

44 Views on the safety of vaccines

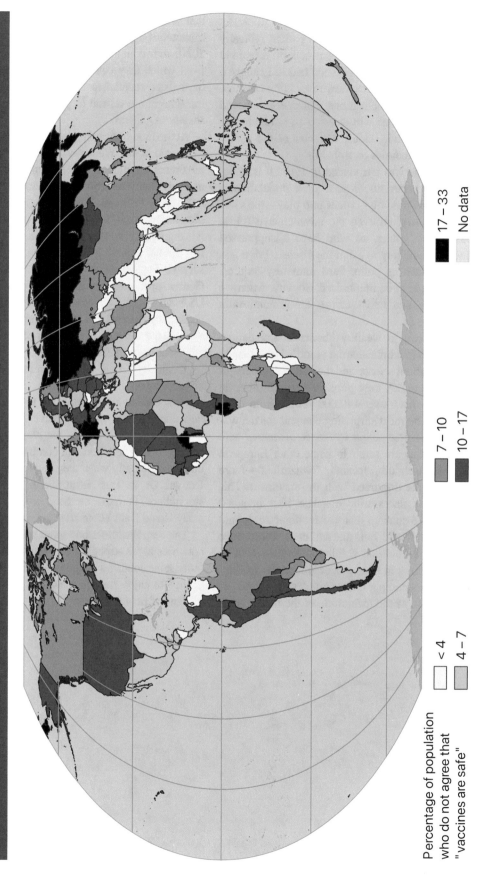

Percentage of population
who do not agree that
"vaccines are safe"

<4
4 – 7
7 – 10
10 – 17
17 – 33
No data

Map 44

45   Good at Maths? (Map 45)

People have a seemingly limitless capacity for ignoring inconvenient truths about the future they are creating—witness the choice of many adults to ignore the warnings of climate scientists. But surely most people would like tomorrow to be better than today? Parents want their children to have access to more opportunities than they themselves experienced. Older citizens hope that younger generations will contribute to a successful society. Politicians (most of them, at least) are keen to ensure their country's workforce has the skills and knowledge necessary for the economy to thrive. Education is key to all this. An educated populace may not be sufficient to guarantee a better future, but it is a necessary starting point. A society whose members are literate, numerate, and capable of logical argument, surely has a better chance of making wise decisions than one whose members are illiterate, innumerate, and prone to unsound reasoning.

In 1997, the Paris-based Organization for Economic Co-operation and Development (OECD), an international body that works with governments and policy makers to establish evidence-based standards and find solutions to social, economic, and environmental challenges, launched its Program for International Student Assessment (PISA). The introduction to the document that announced PISA to the world stated this new framework for assessment would obtain data relating to the following questions:

1. How well are young adults prepared to meet the challenges of the future?
2. Are they able to analyse, reason and communicate their ideas effectively?
3. Do they have the capacity to continue learning throughout life?

The idea behind PISA was that, given the increasingly global nature of the knowledge-based economy, education systems needed to compare their students not just to local or national standards but also to the best-performing schools around the world. By benchmarking student performance against international standards, so the argument went, countries could gauge how prepared their young people would be to participate in a global society. (The PISA developers were prescient in focusing on the global nature of a knowledge-based economy. In 1996, while they were finalising their thinking, Google did not exist; the World Wide Web consisted of just 100,000 websites; and unless you worked in a university your access to the internet was likely to be via a dial-up connection with a speed of 28.8 Kbps. As we now know, technology permits many knowledge workers to offer their services to customers around the globe.)

Starting in 2000, PISA administered a triennial survey of the performance of 15-year-old students in three subjects: mathematics, reading, and science. The latest results to which I have access are those of the 2018 PISA survey, which were released in December 2019. This map focuses on the results for mathematics,

The publication of PISA results often generates a 3-yearly bout of hand-wringing in countries such as the USA: America routinely performs worse than, say, China in the PISA mathematics tests. Worse is the self-congratulation of countries deemed to have 'shot up' the league table: the UK, for example, went from 27th place in mathematics in 2015 to 18th in 2018 and this was seen by politicians as a vindication of the efforts of previous governments.

Are either of these reactions to PISA appropriate? Or healthy?

Before digging into PISA more deeply, let's look at a few of the mathematics questions that schoolchildren are expected to answer. They come from a publicly released question set, which pupils could have encountered in the 2006 survey.

A pizzeria serves two round pizzas of the same thickness in different sizes. The smaller one has a diameter of 30 cm and costs 30 zeds. The larger one has a diameter of 40 cm and costs 40 zeds. Which pizza is better value for money? Show your reasoning.—OECD (2006)

Nick wants to pave the rectangular patio of his new house. The patio has length 5.25 metres and width 3.00 metres. He needs 81 bricks per square metre. Calculate how many bricks Nick needs for the whole patio.—OECD (2006)

In Mei Lin's school, her Science teacher gives tests that are marked out of 100. Mei Lin has an average of 60 marks on her first four Science tests. On the fifth test she got 80 marks. What is the average of Mei Lin's marks in Science after all five tests?—OECD (2006)

These questions highlight the importance of *basic* mathematics in everyday life. The ability to calculate simple ratios, percentages, and probabilities comes in handy all the time. (The relevance of more advanced mathematics in day-to-day activities is less apparent. Since completing my PhD I've seldom had occasion to use the mathematical skills I developed—who needs to calculate Feynman path integrals?) I have seen first-hand how a lack of basic numeracy leads people into poor financial, career, and healthcare decisions. So of course it is important that 15-year-olds possess a certain level of mathematical competence. But what does PISA really tell us about that competence? And does the lack of

45 Good at maths?

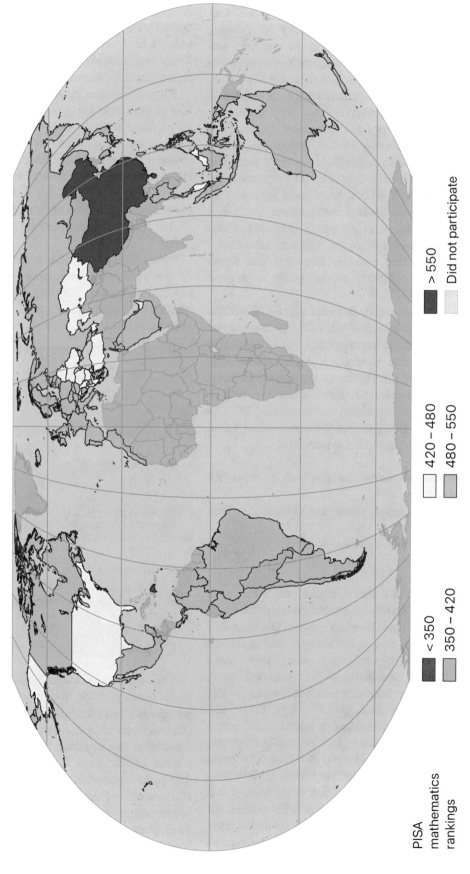

PISA
mathematics
rankings

Map 45

< 350

350 – 420

420 – 480

480 – 550

> 550

Did not participate

mathematical training that is common among politicians, in the UK, at least, lead them to misinterpret PISA league tables and thereby perhaps worsen rather than improve an education system?

Taking that last point first: PISA itself cautions against adopting a narrow interpretation of the league tables. In the 2012 survey, for example, the UK was said to rank between 23 and 31 in mathematics. Few politicians, however, were prepared to deal with that uncertainty. They focused instead on the definite, if unhelpful, statement that the UK was 26th in the maths league table. Poor decisions can flow from a misplaced focus.

The rankings themselves, furthermore, might well be subject to additional statistical uncertainties that are not fully acknowledged. For example, the survey works by having about 4000 children from various schools in each of the participant countries sit a 2-h test. PISA, because it wants to measure a range of skills, creates more questions than a single child can answer—enough questions for a 4.5-h exam—and distributes them between several different test papers. So, by design, not all pupils answer the same set of questions. To compensate for this, PISA use a statistical model to extrapolate from the answers a student gives to get an estimate of how they would have performed had they been given all the other questions. Do you see the problem? In order for this procedure to work you must know the difficulty of each of the questions—but it is not clear that questions have a 'fixed difficulty', particularly when students from different cultures bring different perceptions and experiences to the process of answering questions. Furthermore, the emphasis placed on these tests by parents, teachers, and school authorities differs from country to country. This is likely to cause variations in completion rates—but scores are highly affected by completion rates. So the interpretation of a PISA ranking is more involved than you might think.

I work in education. Given all these uncertainties the emphasis that politicians place on PISA league table position concerns me. It would be silly if the USA, for example, tried to emulate the 'success' of China in mathematics in 2018. It is not just that China's vast population is represented by only four provinces, which creates a source of uncertainty, but that the pedagogic techniques leading to high PISA scores might end up delivering *worse* mathematicians. As computers become more powerful and capable of doing more, the relevant qualities that human mathematicians can bring to bear on problems will be creativity, insight, teamwork—qualities that are not necessarily measured in the PISA mathematics test.

Perhaps even worse than the potential for mistaken policy based on PISA rankings is the view of education perpetuated by PISA. The OECD focused on mathematics, reading, and science—areas that might, I suppose, impact on economic productivity. Now, I am trained as a scientist so I appreciate the importance of these subjects. But what about Arts?

Geography? History? Languages? Literature? Music? Physical education? Social sciences? As educators we should be teaching students to be more than mere economic entities.

Bearing all those caveats in mind, we can say that the top five territories in the 2018 PISA rankings in mathematics were all in the eastern hemisphere. China, or more accurately the four provinces of Beijing, Shanghai, Jiangsu, and Zhejiang, came top with a score of 591. After that came Singapore (569), Macau (558), Hong Kong (551), and Taiwan (531). China, Singapore, and Macau occupied the same top three positions in the other two subjects. In science, the top scores belonged to China (590), Singapore (551), and Macau (544); in reading, the top scores were China (555), Singapore (549), and Macau (525).

There was thus some consistency across subjects in the countries at the top of the PISA rankings. That consistency is reflected across time, too. China is a relative newcomer to the PISA survey, but Singapore came top in mathematics in 2009, 2012, and 2015; Hong Kong came second in those three surveys; Macau and Taiwan placed highly. If one believes that PISA measures something valuable, then the educational systems of these countries appear to be doing something right. Again, though, I would argue that western countries should not take the results of these surveys by themselves as a sign that their own educational systems need to be repaired.

Just as there is consistency at the top of the PISA rankings, so is there consistency at the bottom of the rankings. And I would argue that these figures *do* indicate that change is desirable.

The bottom five territories in the 2018 PISA rankings in mathematics were: Dominican Republic (325), Phillipines (353), Panama (353), Kosovo (366), and Morocco (368). The same countries occupied the same positions in science, and similar positions in reading. These low scores might in part be explained by a lack of experience in administering the PISA survey, since these countries do not have a long history of participation, but they are also likely in part to reflect lower educational attainment.

46   Literacy (Map 46)

My great-great-grandfather, the son of a Norfolk agricultural labourer, moved as a young man to the Victorian boomtown of Middlesbrough. Literacy would have been unimportant to him as he sweated first in the fields and later in the ironworks. Just before he died he married his childhood sweetheart, who travelled with him from Norfolk, and their marriage certificate makes it clear that neither of them were literate. The mental world my forebears inhabited is a place I have difficulty imagining—I read too much to imagine *not* reading— but it's a world in which tens of millions of people still reside.

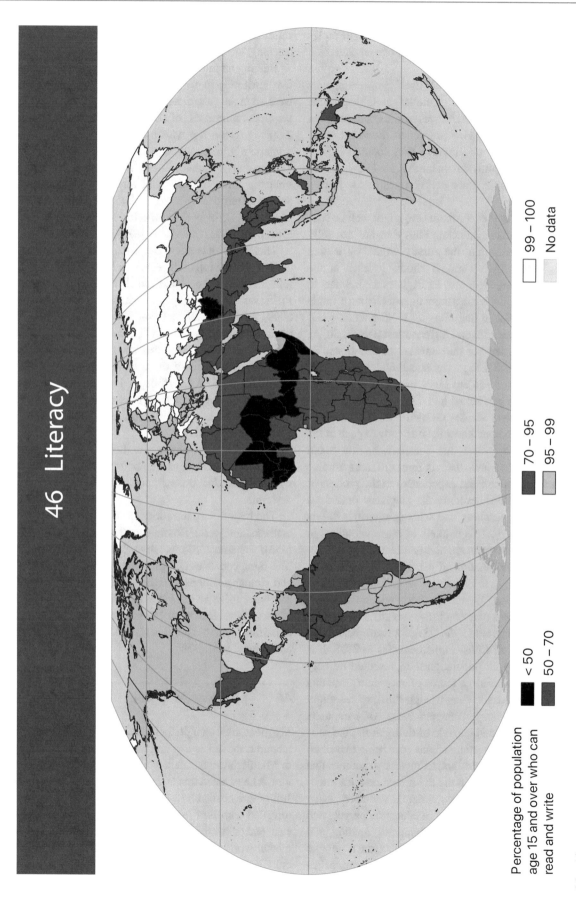

46 Literacy

Percentage of population
age 15 and over who can
read and write

■ < 50
■ 50 – 70
■ 70 – 95
□ 95 – 99
□ 99 – 100
□ No data

Map 46

Country-by-country comparisons of literacy are difficult because of the variability with which rates are measured. In some countries, for example, information on literacy comes from a national census that asks the question: 'Can you read and write?' The question is open to interpretation (some respondents might answer 'Yes' if they can read *or* write; others in the same situation might answer 'No'). If a 'head of household' answers for the entire household then that can further skew the results. Nevertheless, general patterns emerge. Some countries in Europe and Asia claim a literacy rate of 100%; the rest claim a more realistic rate of at least 95%. But in a band of countries across Africa, ranging from Somalia in the east to Liberia in the west, less than half the population is literate. Outside Africa, Afghanistan has the lowest literacy rate: only 38% of citizens are literate, a figure that is likely to decrease as the Taliban impose their rule on the country and limit people's educational opportunities.

The country with by far the lowest literacy rate is Niger. Fewer than one in five people can read and write. Women in Niger fare the worst: not one woman in ten is literate. The reasons for such a poor level of literacy are complex and manifold, but one simple factor is the age profile of the Niger population: the average age is 15, and so the demands of labour are in opposition to those of schooling. The same conflict that led to my ancestor's illiteracy leads to illiteracy in present-day Niger.

> Literacy for all is an integral part of education for all, and both are critical for achieving truly sustainable development for all.—Kofi Annan, Former Secretary-General of UN

In the developed world, of course, we have mandatory schooling. Literacy rates are high. We take for granted the ability to read a page of text. But I wonder whether the *nature* of our literacy is changing?

Much of our reading now takes place on a screen. My research for this book, for example, *had* to happen on screen because I needed to identify and process online data sets. Eyetracking research shows that for screen reading we often use an F-pattern scan, where our eyes fixate mainly at the top and left. In essence, we get information from the top of the screen and then 'dip in' to information on the rest of the screen. When reading a physical book people *used* to employ quite different techniques to extract information from the page. But perhaps our reading techniques are changing. I am unable to point to research on this, but I suspect people carry that F-pattern skimming habit from screen reading over to the reading of novels in traditional book form. Might that in part explain the rise of the audiobook? For people who wish to maintain focus, *listening* to a book can be better than reading it: after all, one can't skim-read an audiobook. (Even there, perhaps, a sense of impatience with the format is emerging: I know of people who play audiobooks at high speed.)

---

47   Marking Time (Map 47)

We live on a clock. Earth's rotation generates the main way in which we mark intervals of time: the day. The orbital period of Earth's satellite gives us another period: the month. The gap between two phases of the Moon (from new to first quarter; first quarter to full; full to last quarter; last quarter to new) gives us a further period: the week. The orbital period of our planet around the Sun gives us yet another interval: the year. Early civilisations made careful astronomical observations, in part because it enabled them to record the passing of time.

A calendar is a method for organising days. The term comes from the Romans, for whom the first day of the month was *calends* from a word meaning 'to call out' after the 'calling out' at the first sight of the new moon. Many calendars are possible—different civilisations have based their calendars on the motion of the Moon, the Sun, and combinations of the two. Unfortunately, any calendar based on astronomical observations is destined to suffer problems. The 7-day week is awkward, for example, because it does not divide evenly into a month; a year of 365 days is not quite right, because on average Earth takes 365 days 6 h 9 min 9 s (and a bit) to complete its orbit of the Sun. Whatever calendar you choose, if you want it to keep in step with the sky then you will need to tweak it from time to time.

In 45 BCE, Julius Caesar reformed the Roman calendar. Back then, the high priests of Rome kept the calendar year in line with the solar year by adding a month when needed. But the system was prone to political influence and it led to confusion. Caesar implemented a calendar he hoped would need no human intervention in order to stay aligned with the Sun. The Julian calendar, with its system of leap years, was a distinct improvement on the Roman calendar; but it was wrong, compared to the stars, by just under 1 day in a century. After 1600 years the discrepancy had built up. The date of the spring equinox, which is important to the Catholic church, was noticeably wrong. In 1582, Pope Gregory XIII took corrective action. He reset the date of the equinoxes and tweaked the Julian calendar by adding an extra rule about leap years: if the year is divisible by 100 and not divisible by 400 then the leap year is skipped. Most of the world has settled on the Gregorian calendar, and we seldom give it a second thought, but there is nothing inevitable about it. Other calendars are possible.

Some countries use a modified version of the Gregorian calendar. Japan, for example, uses the Gregorian calendar but designates the year according to the reign of the current Emperor. Thailand adopted a version of the Gregorian

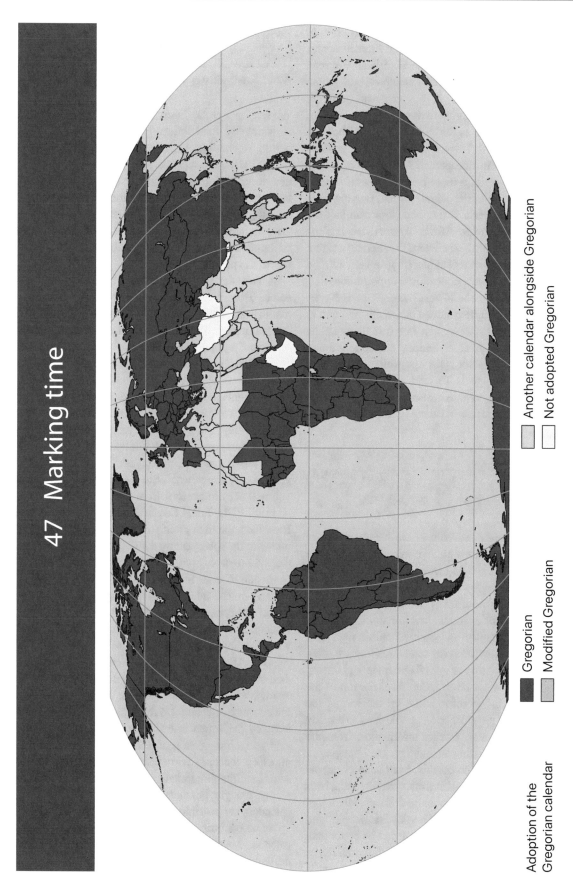

47   Marking time

Adoption of the
Gregorian calendar

- Gregorian
- Modified Gregorian
- Another calendar alongside Gregorian
- Not adopted Gregorian

Map 47

calendar but chose a starting year based on the death of the Buddha, taken to be 543 BCE. North Korea modified the Gregorian calendar so that year 1 celebrated the birth in 1912 of Kim Il-sung, the country's founder. Many other countries use the Gregorian calendar for official purposes, but have a traditional calendar that influences dates of national holidays. China is the obvious example: the date of the Chinese New Year, and other festivals, is based on a traditional lunisolar calendar. Since Gregorian is the official calendar in China however, the map marks it as such.

Many Arabic countries use the Gregorian calendar for civil purposes and run the lunar Hijri calendar alongside it to determine the dates of religious festivals.

Of more interest are the handful of countries that have not adopted any aspect of the Gregorian calendar. Iran and Afghanistan use the solar Jalali calendar, in which the new year begins on the vernal equinox as determined by astronomical observations from Tehran. Ethiopia uses the solar Ge'ez calendar, which has 12 months of 30 days and a 5- or 6-day intercalary month at the end of the year. Bhutan uses a version of the lunisolar Tibetan calendar, with a year consisting of 12 or 13 lunar months. Nepal uses the Hindu lunisolar calendar.

We live on a clock—but cultures differ in how to use it as a timepiece.

48   UFO Sightings (Map 48)

The National UFO Reporting Center (NUFORC), an organisation based in the state of Washington, USA, maintains a database of well over 100,000 UFO sightings. The database possesses a number fields—location, duration, and date of sighting; the shape of the UFO; eyewitness details. The internal consistency of the data leaves much to be desired but, following a data cleansing exercise, it is possible to construct a map showing the location of more than 80,000 individual UFO sightings.

The most readily apparent feature of the map shown here is that sightings predominantly occur in North America and Western Europe. Why should that be? (In looking for an answer I assume UFOs are not visitations from aliens obsessed with western civilisation.)

One explanation for this skewed distribution could be that, since NUFORC is based in the USA, American reports of strange lights in the sky are more likely to be recorded. Good communication links between Western Europe and the USA mean that European sightings are also likely to be recorded. In other countries there might be no mechanism for notifying national authorities, much less NUFORC, of unexplained atmospheric phenomenon. So perhaps this distribution of sightings is what one might expect. On the other hand,

cultural bias could well play a role in explaining this uneven distribution of UFOs.

The first 'modern' UFO sighting occurred on 24 June 1947, when a private pilot called Kenneth Arnold reported seeing a line of nine cigar-shaped objects flying past Mount Rainier. The American press, building on Arnold's description that the objects moved like skimmed saucers, came up with the evocative and mysterious term of 'flying saucers'. Hundreds of sightings quickly followed, the most famous being the so-called Roswell incident—the crash of a weather balloon in July 1947, which led to numerous conspiracy theories involving flying saucers. The American people became sensitised. Lights in the sky, whether mis-sightings of Venus or weather balloons, momentary glimpses of aircraft or meteorites, were reported as UFOs. And the burgeoning popularity of SF films meant people were more than willing to identify these unidentified objects: in the minds of many people, UFOs were flying saucers were alien spaceships. Western Europeans fell under much the same influence.

People report hundreds of UFOs every year. Often, upon investigation, the object can be identified: the UFO becomes an IFO—perhaps Venus, or a balloon, or a hoax, or the result of unusual atmospheric conditions. Some remain unidentified. Interestingly, the yearly IFO/UFO ratio remains roughly constant regardless of how many UFOs are reported—which is what one would expect from random misperception and misreporting.

In recent years UFOs have undergone something of a rebranding exercise. A term often used today is 'unidentified aerial phenomenon' (UAP). In July 2021, the Pentagon published an assessment based on 144 UAP reports made by military personnel. One UAP report had an explanation (it was a deflated balloon); the others remained unidentified. Clearly, it is in everyone's interest for the military to identify these UAPs: it is easy to imagine what might happen if a jet fighter erroneously fires a missile at a UAP. But it seems to me the scientific identification of UAPs is not a task for those of us interested in the possibility of extraterrestrial intelligence. Technical specialists from other fields—aeronautics, psychology, optics, remote sensing, sociology, software engineering, vision science—should all be involved before astrobiologists are asked to investigate UAPs.

If you have never seen a UFO then you have not been looking long enough nor closely enough at the sky. I have seen a UFO, and it remains a vivid memory. But was it an alien spacecraft? No, it was a UFO—an object in the sky that I am unable to identify.

48 UFO sightings

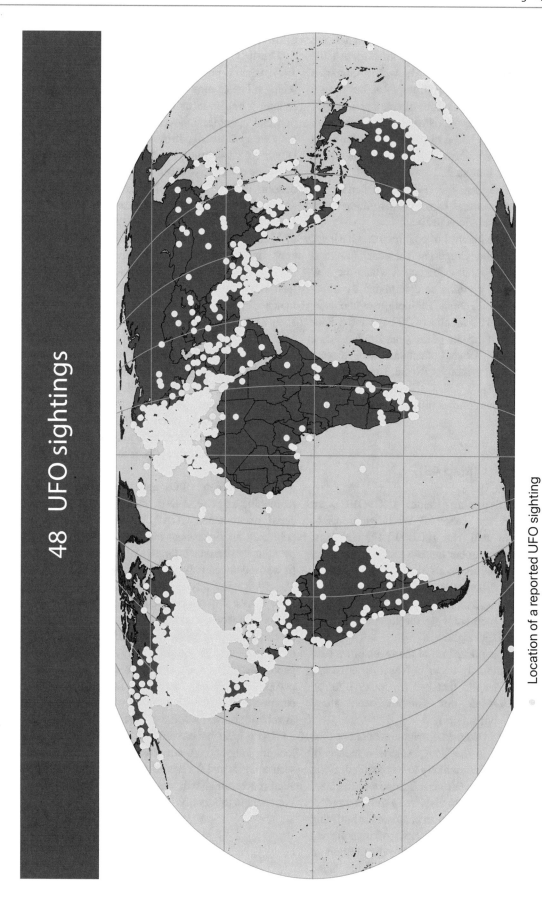

● Location of a reported UFO sighting

Map 48

49   How the World Believes (Map 49)

Religious belief presumably has its roots in the way our ancestors saw the world. It would make evolutionary sense if early humans perceived agency in things. For example, an early human's survival required an understanding of the fact that a lion has agency. When hunter-gatherers noticed a rustling sound in grass, it was safer for them to vamoose rather than cogitate on whether the cause of the rustling was wind or lion. Humans with this understanding and a facility for rapid decision-making tended to live longer, have more children, and pass on those traits. But then humans started attributing agency to things that really don't possess agency—rain, stars, thunder. And they began applying their 'theory of mind'—a remarkable facility that enabled them to gauge other people's intentions: whether a stranger wanted to trade or steal; whether it was sensible to share food or prepare to fight—on those things that have no agency. When early humans began trying to gauge the intentions of rain or stars or thunder ... religion was on its way.

Religion still has a hold on much of the world's population. The Pew Research Center, established in 2004, conducts public opinion polling, demographic research, content analysis, and various other types of data-driven social science research. As one element of its social science research activity, the Center investigates attitudes to matters of faith and religion. According to the Pew database, in only seven countries is 'unaffiliated' the predominant religious classification (and even in those countries non-affiliation does not necessarily imply atheism or agnosticism).

Christianity is the most geographically widespread faith, being the predominant religion in 162 countries. Christianity has particular strongholds in South Africa, South America, and Southern Europe (and the Vatican, of course). Many countries are nominally Christian, but in several European countries the percentage of believers is in the range 60–70%.

Islam is the predominant faith in 52 countries. As is clear from the map, countries in North Africa and the Middle East are particularly devout.

Compared to Christianity and Islam, other religions are more geographically localised. Those religions, however, can still have a large number of believers. Hinduism is the major faith in only three countries: India, Mauritius, and Nepal. The population of Mauritius is only 1.2 million, less than half of whom are Hindus, so the total number of believers there is relatively small. The population of Nepal is just under 30 million, more than 80% of whom are Hindus, so a rather larger number of believers live there. But it is India's vast population that accounts for the popularity of Hinduism: 80% of India's 1.38 billion people are classed as Hindus. If the Pew database is correct, the Hindu faith is carried by more than a billion people.

Buddhism is the predominant religion in eight countries, with Cambodia and Thailand being particularly devout, Myanmar slightly less so. In Singapore the proportion of Buddhists is larger than other beliefs, but many faiths are represented in this country: only a third of Singaporeans are Buddhists.

The diaspora means Jews live in many countries, of course, but only in Israel is Judaism the dominant faith.

Macau, Taiwan, and Vietnam are interesting: in these places native folk religions remain popular. In Vietnam, the ethnic folk religion is aligned to a set of local worship traditions. The 'spirits' or 'gods' involved in worship are deities related to the natural world, to community, and to ancestors. In Taiwan, the folk religion is a combination of Buddhism and Taoism, with elements of Confucianism.

49 How the world believes

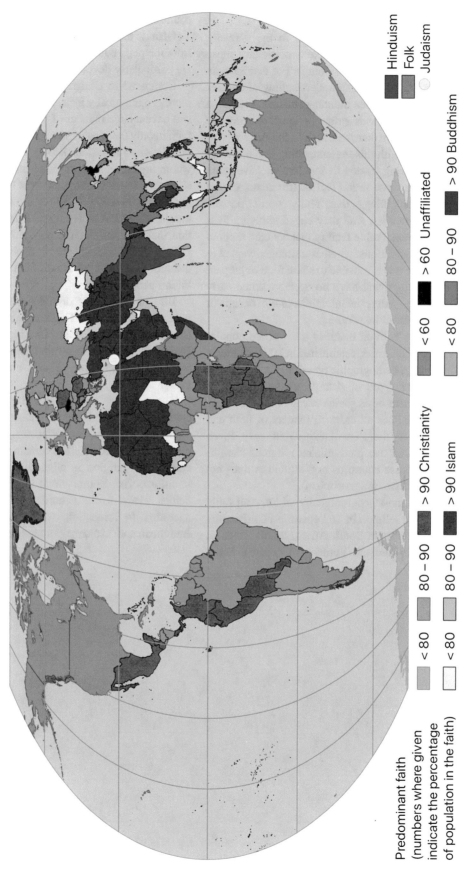

Predominant faith
(numbers where given
indicate the percentage
of population in the faith)

< 80   80 – 90   > 90 Christianity
< 80   80 – 90   > 90 Islam

< 60 Unaffiliated
< 80   80 – 90   > 90 Buddhism

Hinduism
Folk
Judaism

Map 49

50   The Happiness of Nations (Map 50)

As we shall see in Map 74, economists often use gross domestic product (GDP) as a measure of the health of a national economy. But the definition of GDP contains within it several incongruities. For example, an act of marriage can reduce GDP: pay a housekeeper and you add to your nation's GDP; marry your housekeeper and, if the same service is done for 'free', then the work does not count for GDP purposes. That being the case, does GDP make much sense as a measure? The inaptly named SIN countries—Scotland, Iceland, New Zealand—are now trying to include indicators of wellbeing when developing and assessing government policy. But what sort of indicators to use? Gross national happiness, perhaps?

> I have learned to seek my happiness by limiting my desires, rather than in attempting to satisfy them.—John Stuart Mill

Although I am sceptical about measuring happiness (I am not sure when I myself am 'happy'), this is a well-established field: at least one academic journal—*Journal of Happiness Studies*—is devoted to the subject. And if it can be measured then we can compare countries in terms of the happiness of their citizens. Since 2012, the UN Sustainable Development Solutions Network have published the *World Happiness Report*, an annual ranking of national happiness based on responses to the World Gallup Poll. Survey participants are asked to imagine a ladder in which the best possible life they could have is a 10 and the worst possible life is a 0. They then rate their current life on that scale. The hypothetical country of Dystopia has values equal to the world's lowest national

averages for each of six factors: GDP; life expectancy; generosity; social support; freedom; and corruption. Countries are then ranked against Dystopia. The latest report to which I have access gives happiness scores averaged over the 3 years 2017–2019.

Finland (score 7.809) was for the third year the happiest country in the world. Afghanistan (2.567) was the unhappiest. (Three decimal places, as given in the *Report*, is surely a spurious level of precision: can Montenegro and Russia, both 5.546, *really* be equally happy places?)

> You will never be happy if you continue to search for what happiness consists of. You will never live if you are looking for the meaning of life.—Albert Camus

A glance at the map of happiness shows few surprises. India and much of Africa are low on the happiness scale; northern Europe, Canada, Australia, and New Zealand score more highly. But how much meaning can we assign to 'happiness' here? Consider, for example, Finland. It always scores highly. But these polls could be self-fulfilling. Perhaps Finns feel a duty to respond 'happily' because polls show Finland is a 'happy' place? In Map 52 we discuss suicide—and it turns out Finns are over four times more likely than South Sudanese (citizens of the second most unhappy country, score 2.817) to kill themselves. Broad statistics can often gloss over the nuance inherent in these difficult concepts.

Happiness as defined in the *Report* is perhaps no better than GDP as a full measure of a nation's economy. But surely, after a pandemic, we should take *some* account of wellbeing?

50 The happiness of nations

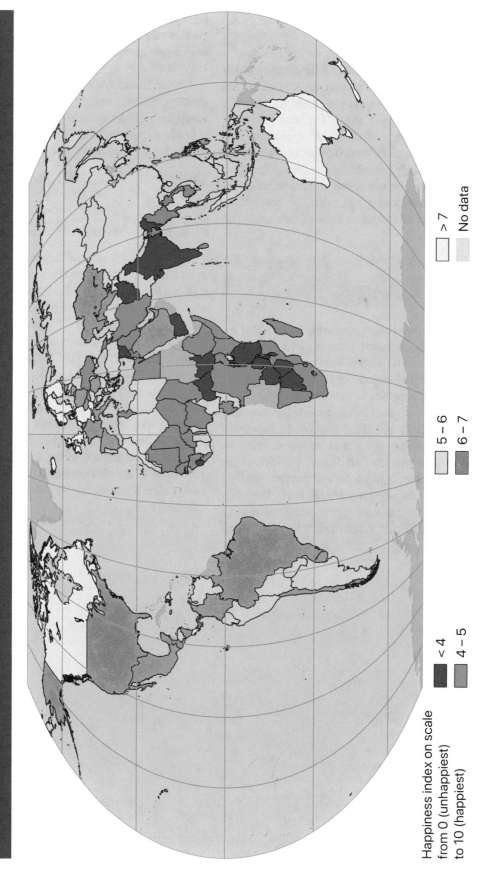

Happiness index on scale
from 0 (unhappiest)
to 10 (happiest)

< 4
4 – 5
5 – 6
6 – 7
> 7
No data

Map 50

The World of Wellbeing

SARS-CoV-2, the virus responsible for the disease Covid-19, did not exist when I began researching these maps. Then the blind workings of evolution created a tiny package of genetic material with the capacity to upend our everyday existence. As I write, the illness has claimed more than 5.6 million lives. Our response to the pandemic created its own problems: lockdowns and the curtailment of economic activity caused mental distress and physical harm. This new pandemic illness, furthermore, has been visited upon a world suffering from existing inequalities of health, wealth, and security. The full impact of Covid-19 will not become clear for several years, however, so this chapter considers wellbeing as it was just before the start of the pandemic.

Of course, health and wellbeing depend on more factors than just infectious disease. So the maps in this chapter illustrate broader aspects of wellbeing, including the experience of violence, pain, and mishap. When contemplating all this it is easy to be downhearted. So it is important to note that, for those lucky enough to live in certain countries, life expectancy at birth—particularly for women—is higher than ever.

51   Homicide (Map 51)

All societies condemn murder. The crime, by definition, causes the loss of human life. But its effects ripple out beyond the victim: murder ruins the lives of the victim's family, damages the lives of the victim's friends and colleagues, and casts a shadow over society in general. Because the crime of murder is so grave, and because it is so specific, homicide statistics are relatively easy to collect. Data for the crime of homicide are more reliable and valid than, say, data for financial fraud.

The United Nations Office on Drugs and Crime (UNODC) define 'intentional homicide' as a crime with the following three elements:

1. The killing of a person by another person (an *objective* element).
2. The intent of the perpetrator to kill or seriously injure the victim (a *subjective* element).
3. The unlawfulness of the killing (a *legal* element).

If a killing has those three elements then UNODC label the act as an 'intentional homicide'. UNODC apply that label regardless of how particular national legislation might classify the crime (which is why the UNODC figures sometimes differ from those you might see elsewhere). This definition enables UNODC to disentangle intentional homicide from other forms of killing, such as war, suicide, justifiable homicide, or manslaughter caused by recklessness or negligence, and thus provide a fair comparison between nations. Note that killings caused by terrorism are, under this definition, to be classed as acts of intentional homicide.

The data in the map shown here, the latest data on the topic to which I have access, come from the UNODC 2019 report entitled *Global Study on Homicide*. (The report is published as six booklets: an executive summary; an overview of international homicide rates; an examination of drivers and mechanisms of homicide; an analysis of the relationship between homicide and sustainable development; an overview of gender-related homicide; and a survey of child homicide.) The majority of countries have data for 2017, with data for most remaining countries referring to 2016 or 2015. It hardly needs pointing out that we should not assume countries for which no data is available are havens of peace and tranquillity: some of these countries lack relevant statistics because they lack effective systems of government, and violence is hardly uncommon in such states.

If we take homicide rate as a proxy for violence in general then the map shows that some countries are more peaceful than others.

The homicide rate in several Asian countries—Japan, Singapore, China, Indonesia, South Korea—is below 1 per 100,000. (For Singapore and Japan the 2017 homicide rate

© The Author(s), under exclusive license to Springer Nature Switzerland AG 2023
S. Webb, *Around the World in 80 Ways*, https://doi.org/10.1007/978-3-031-02440-5_7

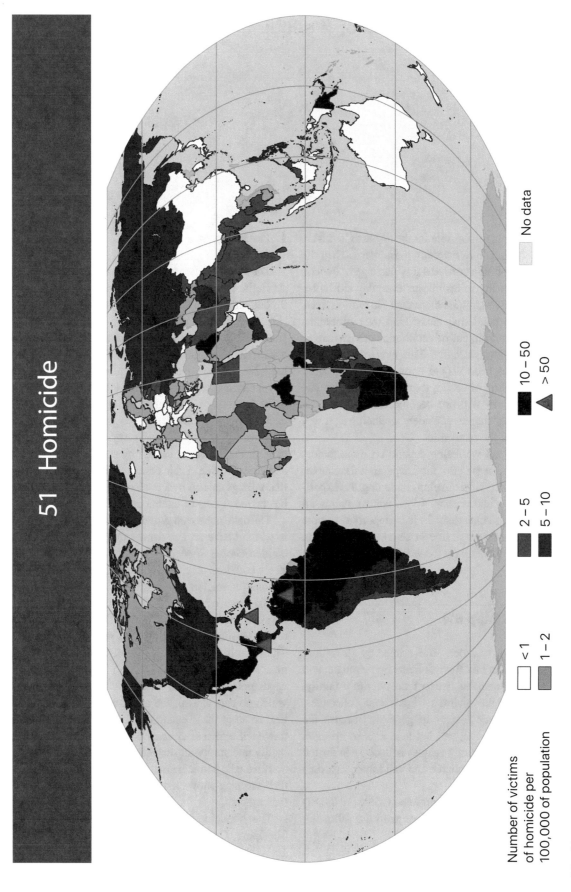

51 Homicide

Number of victims
of homicide per
100,000 of population

<1

1 – 2

2 – 5

5 – 10

10 – 50

> 50

No data

Map 51

Fig. 40 A busy market in Santa Ana, El Salvador. The people here, as in other parts of El Salvador, live under the daily threat of violence and extortion. Santa Ana, the country's second largest city, is home to street gangs and is a transit point for criminals smuggling drugs and various other forms of contraband between El Salvador and Guatemala. (Randal Sheppard, CC BY-SA 2.0)

was 1 per 500,000.) The homicide rate in several European and Nordic countries—Luxembourg, Norway, Switzerland, Cyprus, Czechia, Greece, Portugal, Spain, Netherlands, Poland, Iceland, Ireland, Slovenia—is also less than 1 per 100,000. (In Liechtenstein in 2016, the latest year for which figures are available for that small country, the homicide rate was 0.) As evidenced by the homicide rate, a number of other countries—Oman, UAE, New Zealand, Australia—are peaceful places.

In the rest of western Europe, in Canada, and in a few other countries, the homicide rate hovers between 1 and 2 per 100,000 of population.

And then there are the countries where violence is the background noise to daily life.

Three countries endure homicide rates in excess of 50 per 100,000. These countries, marked on the map by red triangles, are El Salvador (where the homicide rate in 2017 was 61.8 per 100,000 of population), Jamaica (57 per 100,000), and Venezuela (56.3 per 100,000).

The citizens of several countries in Central America endure high levels of violence, a situation that undoubtedly contributed to the increase in the number of refugees congregating on the USA's southern border during 2018–2019. We have already noted that El Salvador has the highest homicide rate of all, but Honduras, with a homicide rate of 41.7 per 100,000, and Belize, with a rate of 37.9, are not far behind. Nevertheless, regional variations exist. A £40 bus ride will take you from Santa Ana in El Salvador (Fig. 40) to Managua in Nicaragua, but the homicide rate in the former country is more than eight times that of the latter.

One sees clusters within countries, too. Colombia, for example, has a high homicide rate of 24.9. (This is admittedly less than half the rate of bordering Venezuela, but that is a low bar.) Residents of different parts of Colombia, however, experience life differently: some areas are peaceful while the capital Bogotá can be a dangerous place to live. Even within Bogotá people's experiences differ: it is those parts of town on 'the wrong side of the tracks' that suffer the violence.

The UNODC report makes it clear that criminal activity takes as big a toll on life as armed conflict. Consider homicides just by *organised* crime: in the period 2000–2017 organised crime caused about 1 million homicides—roughly the same number of lives lost in armed conflicts over the same period.

The same report also offers fascinating if grim reading about the forms of homicide. Across the world, for example,

Fig. 41 Left: The Singapore skyline (Unwicked, CC BY-SA 4.0). Right: A street in downtown Kingston, Jamaica (Christina Xu, CC BY-SA 2.0). Both places experienced similar colonial histories in the late nineteenth and early twentieth centuries, but in recent decades their paths have diverged. In 2017, the homicide rate in Jamaica was 285 times that of Singapore, with gang violence and shootings being common

more than half of all homicide victims were killed by firearms. In the Americas, about 75% of homicides involved firearms. There seems to be a correlation here: countries with high levels of firearm possession are often countries with high rates of homicide. Countries in which knives and sharp objects are the main murder weapon tend to have low rates of homicide. Perhaps this is unsurprising: a person in possession of an assault rifle can kill more people more quickly than a killer in possession of a knife. Correlation does not imply causation, but surely it is worth asking the question of an electorate: would you not feel safer if gun ownership were restricted?

The report also offers insight into victims and perpetrators. Across the globe, 81% of victims of recorded homicide were men and boys. In the Americas, the rate of male homicide was about ten times higher than the rate of female homicide. Males everywhere are much more likely than females to be the victim of murder (although women and girls are much more likely than men and boys to be killed by family members and intimate partners).

The difference between male and female is even starker when it comes to the perpetrators of homicide. Across the globe, men commit about 90% of all homicide.

So much for the data. Is it possible to understand *why* homicide occurs and thus, perhaps, help prevent it?

In thirteenth century Western Europe homicide rates were as high as the most violent places in the world today. For more than 600 years those rates have been on a downward trend. One can propose many possible explanations for this welcome fact. Murder rates fell as state powers expanded, and perhaps the latter helped cause the former; the rule of law increased over that time and presumably that too had an

effect; and of course people became better educated over that period, a change that may also have played a role. It is not clear which of these interventions might have been most effective. Perhaps it was a combination of things.

In this respect it is interesting to look at countries with similar histories but different patterns of homicide. Consider Singapore and Jamaica (Fig. 41). Both were British colonies; both had similar homicide rates in the nineteenth century, which fell in line with trends in western Europe; both had rates that surged in the 1920s; but then from about 1950 they took different paths. The homicide rate in Singapore dropped and continued to drop, so it is now one of the lowest on Earth. In Jamaica the homicide rate continued to rise, to a level that is now higher than almost all other countries. Singapore enjoyed rising levels of education, life expectancy, and wealth. Jamaica suffered rising levels of gun crime, gang violence, and poverty. Why?

If sociologists could answer this, politicians might be able to implement policies to reduce untold amounts of suffering.

52   Suicide (Map 52)

In 2019, more than 700,000 people died due to suicide. Globally, suicide was the fourth leading cause of death in people aged 15–19.

Every country has people who choose to kill themselves but several factors combine to make a country-by-country comparison of rates difficult. Different countries have differing criteria for reporting suicide; they have varying ways of ascertaining people's intention of taking their own lives; they have different officers responsible for completing death

52 Suicide

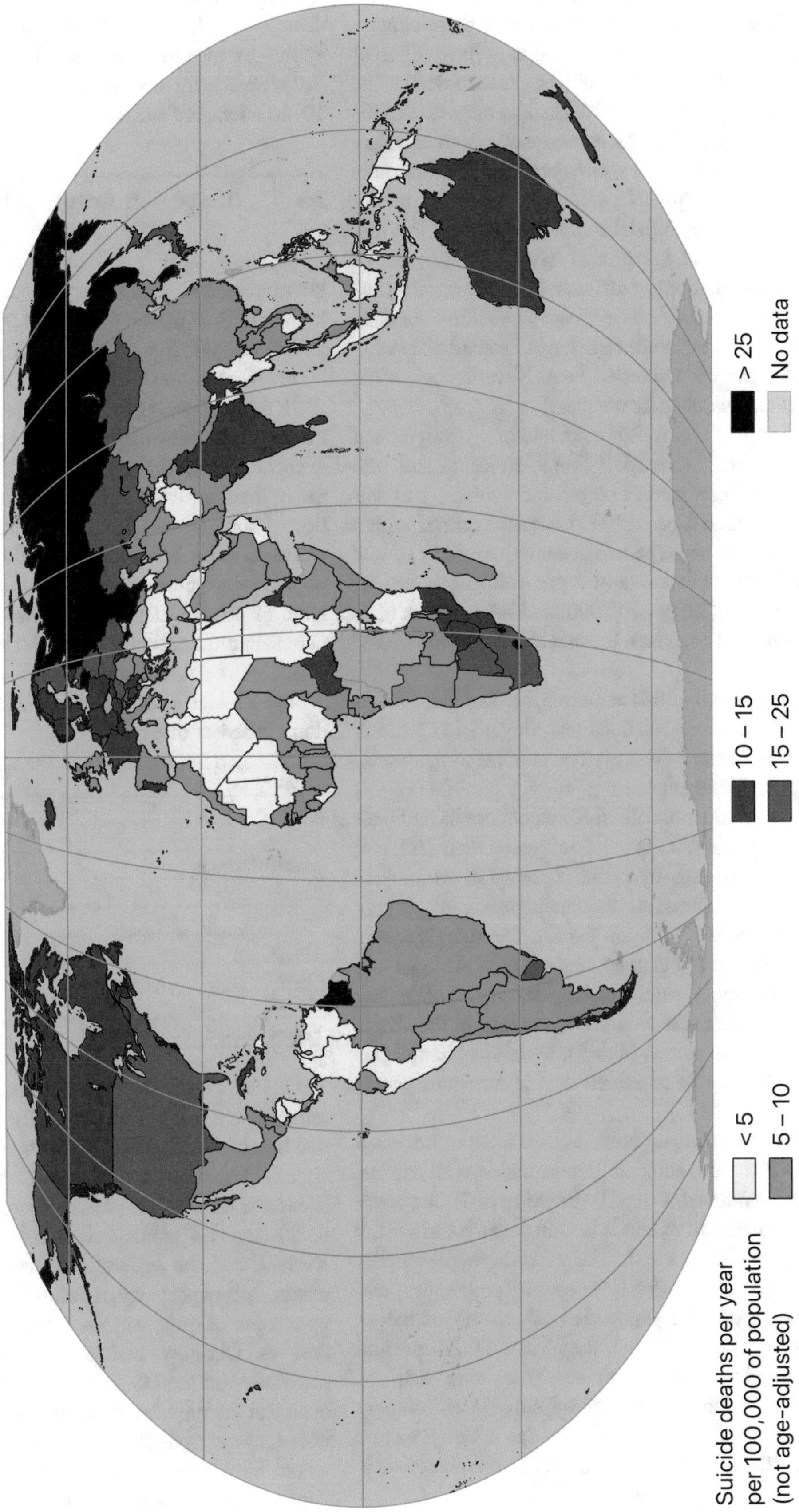

Suicide deaths per year
per 100,000 of population
(not age-adjusted)

< 5
5 – 10
10 – 15
15 – 25
> 25
No data

Map 52

certificates; they have different attitudes to the need for carrying out forensic investigations; they may permit euthanasia, or physician-assisted suicide; and they have different expectations regarding confidentiality of cause of death. The World Health Organization's Global Health Estimates (GHE) use standard categories, definitions, and methods to calculate rates that permit cross-country comparisons; the map here relies on these rates. The GHE estimates, however, are not necessarily the same as official national estimates. Furthermore, because different age groups have different rates of suicide and the populations of different countries have different age structures, it makes sense to age-standardise the data. The map here shows the crude rather age-standardised statistics. Therefore caution is needed here. Nevertheless, some differences between countries are clear.

The GHE reports that in 2019 six nations—Antigua and Barbuda; Barbados; Grenada; Saint Vincent and the Grenadines; São Tomé and Príncipe; and Jordan—had suicide rates of less than 2 per 100,000. Some countries with a troubled recent past have, perhaps counterintuitively, a low suicide rate. Syria, with a rate of 2 per 100,000, is not far behind Jordan. Iraq (3.6 per 100,000) and Afghanistan (4.1) fare better than most countries. It is not clear why this should be the case.

At the opposite end of this sombre list is Lesotho (with a suicide rate of 72.4 per 100,000) and Guyana (31). South Korea, which has one of the largest economies in the world, is fourth on the list (a suicide rate of 28.6 per 100,000). It appears that some older people in Korea choose to end their life early rather become a financial burden on their children; and many students take their life if they fail to achieve academic success. Lithuania, another high-income, highly developed country, is seventh on the list. The suicide rate in Lithuania (26.1) is more than 65 times that of Antigua and Barbuda—a difference surely calling out for explanation. (Note that Lithuania ranks well above average in the *World Happiness Report*, as seen in Map 50. If 'national happiness' is something that can be measured then it correlates poorly with the act of suicide.)

The World Health Organisation collects more information about suicide than pure numbers. The variation with age has already been mentioned. Overall, people over 70 are more likely to kill themselves, except in certain countries where the 15–30 age group is most at risk. The methods people employ to take their own lives differ from country to country, and probably reflect ease of access. In the USA, more than half of suicides involve guns. In China, the most common method involves poisoning from pesticides (see Map 28). In Singapore, Japan, Hong Kong, and South Korea jumping from a height is common (the Mapo Bridge in Seoul gained the nickname 'Bridge of Death'). Another common method is by carbon monoxide poisoning.

Different methods often lead to different outcomes. The death rate with guns is 80–90%; shooting oneself is effective. With drug overdose, the death rate is 1.5–4%.

One last grim thought: for every recorded suicide there are 10–40 attempted suicides.

53   Obesity in Adults (Map 53)

A person's body mass index (BMI) is their weight in kilograms divided by the square of their height as measured in metres. So a person of height 1.7 m who weighs 70 kg has a BMI of $70/(1.7 \times 1.7) = 24.22$. An adult with a BMI of 30 or above is considered obese. One can argue the merits of BMI, particularly when it comes to the health of individuals. Professional rugby players, for example, often have a BMI in excess of 30 but their high body weight derives from muscle rather than fat; these individuals are extremely fit. The location of body fat also seems to be important: abdominal fat is worse than fat on hips and thighs. On the scale of entire populations, however, BMI provides a quick and easy estimate of obesity. This map shows the percentage of obese people in different countries in the year 2016. So let's look in a little more detail at obesity rates around the world.

The ten states with the highest share of the adult population classed as obese are:

Nauru	61% (population 11,200)
Cook Islands	55.9% (population 17,459)
Palau	55.3% (population 21,503)
Marshall Islands	52.9% (population 53,066)
Tuvalu	51.6% (population 11,192)
Niue	50% (population 1624)
Tonga	48.2% (population 100,651)
Samoa	47.3% (population 195,843)
Kiribati	46% (population 110,136)
Federated States of Micronesia	45.8% (population 104,937)

One reason for the high obesity rates in these places is a move from a diet based on fish, fresh fruit, and vegetables to one based on calorie-dense imported food. This change has already impacted the health of the inhabitants of these islands: a third of people on Nauru are diabetic; vitamin A deficiency is a public health risk in the Marshall Islands, Kiribati, and the Federated States of Micronesia; in some places, life expectancy is falling. Of course, the combined population of these nations is small, about the same population as Glasgow in Scotland, Essen in Germany, or Milwaukee in the US. In global terms these are small numbers. But the prevalence of obesity is increasing in countries with far larger populations. Fifty years ago, this world map would have been a mix of light shades. Now, deep shades dominate. If current trends continue, then by 2030 almost half

53 Obesity in adults

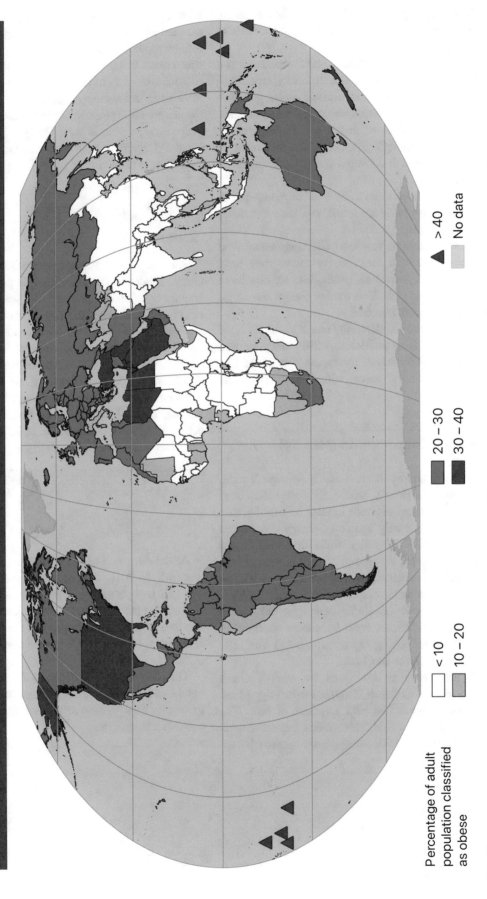

Percentage of adult
population classified
as obese

☐ <10	▲ > 40
▨ 10 – 20	☐ No data
▨ 20 – 30	
▨ 30 – 40	

Map 53

of the world's adult population will be overweight or obese. And that has implications for the future health of hundreds of millions of people.

The US Centers for Disease Control and Prevention lists a range of health conditions associated with obesity: type 2 diabetes; hypertension; coronary heart disease; stroke; gallbladder disease; osteoarthritis; sleep apnoea; and some cancers. The burden of these diseases is not just a problem for individuals but also for policy makers. The treatment of obesity-related illness will consume increasing amounts of public resource.

Several factors appear to contribute to this pandemic of obesity. Calorie-dense foods are readily available and easily affordable for increasing numbers of consumers. Those foods, moreover, are often marketed with aggression. And as people have increased their energy intake they have reduced their energy expenditure: populations are now more sedentary than in the past. That excess energy has to go somewhere. Bodies store it as fat.

Even Vietnam, whose adult obesity rate of just 2.1% is the lowest in the world, faces problems. A 2017 study showed that a move to a western way of life means 30% of children are obese. The obesity wave will soon crash over Vietnam too.

54   Alcohol Consumption (Map 54)

People have been getting drunk—squiffy, pie-eyed, tight, inebriated, plastered—since ... well, since before they were people. Our distant ancestors used to feast on fruit because fruit sugars provide an easy source of energy. Fruits also contain yeast. In over-ripe fallen fruits the yeast breaks down sugar into ethanol, the type of alcohol present in beer and wine. So natural selection provided our ancestors with an enzyme that enabled them to metabolise alcohol. Modern humans can thus process alcoholic beverages, but so can our cousins the chimpanzees (who have been observed to get sozzled on the fermenting sap of palm trees). Our relationship with alcohol goes *way* back.

In 2018, archaeologists discovered the world's oldest brewery: they found a cave near Haifa, in modern-day Israel, containing stone mortars that held a residue of beer. The mortars were 13,000 years old. This ancient fermented brew was probably more akin to alcoholic porridge than to what we

know as beer (these were stone age people, after all, and were thus unlikely to be observing anything along the lines of the Reinheitsgebot, or German Beer Purity Law). Nevertheless, it was a drink that presumably served an important purpose.

Archaeologists have also detected traces of a fermented drink from jars made by people living in modern-day China; the jars are 9000 years old. People were producing wine 8000 years ago, in what is now Georgia. The Babylonians and ancient Egyptians brewed alcoholic drinks. Historically, then, people have always boozed.

> I have taken more out of alcohol than alcohol has taken out of me.—Sir Winston Churchill

And what about the modern consumption of alcohol? The Global Health Observatory provide per capita figures, on a country-by-country basis, for the total amount of alcohol consumed (this being the sum of recorded and unrecorded alcohol) per person (15 years of age or older) over a calendar year, in litres of pure alcohol, adjusted for tourist consumption. As the map shows, the citizens of many countries in North Africa and the Middle East have low levels of alcohol consumption. Some countries are essentially dry: Bangladesh, Kuwait, Libya, Mauritania, and Somalia all registered zero per capita alcohol consumption in 2016. Religious teaching and observance presumably plays the major role here.

> When I was younger I made it a rule never to take strong drink before lunch. It is now my rule never to do so before breakfast.—Sir Winston Churchill

At the other end of the scale one finds mainly European countries. In 2016, the per capita consumption in Moldova was 15.2 L of pure alcohol; the Moldovans were followed by Lithuanians (15 L), Czechs (14.4 L), and Germans (13.4 L). The only non-European country in the list of most inebriated nations was Nigeria, which has the same consumption rate as Germany.

The citizens of many countries treat alcohol consumption lightly, as I myself have done in the paragraphs above. But in some countries, excessive alcohol consumption is a major public health concern.

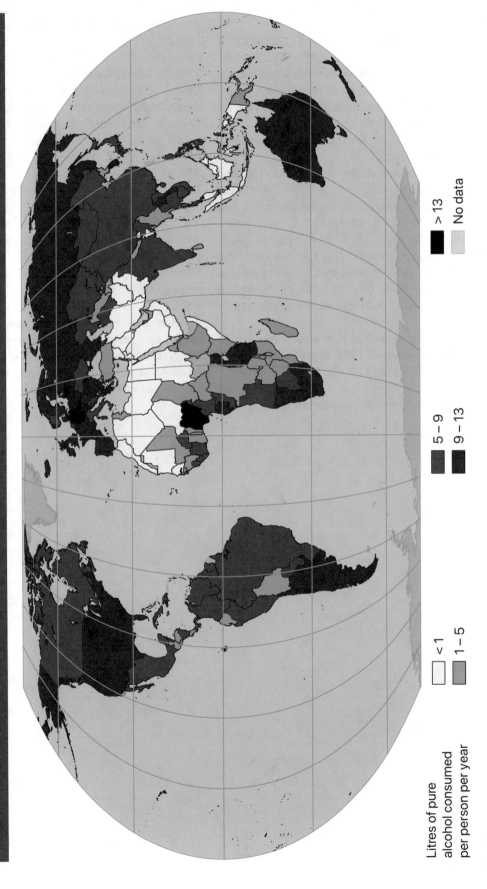

54 Alcohol consumption

Litres of pure
alcohol consumed
per person per year

<1
1 – 5
5 – 9
9 – 13
> 13
No data

Map 54

55   Road Traffic Deaths (Map 55)

According to the World Health Organisation's *Global Status Report on Road Safety 2018* the year 2016 saw 1.35 million deaths caused by road traffic injuries. More than half of those deaths were of 'vulnerable' road users: pedestrians, cyclists, and motorcyclists. This was the eighth leading cause of death among all age groups. For those under the age of 29, it was the leading cause of death. But the worldwide distribution of deaths was far from uniform.

In tiny San Marino no one died on the road. The road traffic fatality rate in most of Europe was below 5 deaths per 100,000 of population. Singapore, Japan, and Israel had similar low rates. In Africa, though, roads were lethal. Liberia's roads were worst (35.9 deaths per 100,000); Fig. 42 shows a typical example of a Liberian road. But many other African countries were not far behind, despite having proportionally fewer cars than Europe. In African countries, deaths of cyclists and pedestrians were particularly numerous.

Non-African countries with a poor road safety record included Dominican Republic (34.6 deaths per 100,000), Venezuela (33.7), and Thailand (32.7).

Why is it safe to be on the roads of some countries and not others? The WHO report, which looks at how to make roads and vehicles safer, and how legislation might influence road user behaviour, addresses this question.

In large part the answer lies in whether a country has laws that address five key risk factors. The first is *speed*: the faster a vehicle travels the more likely the risk of a crash and the greater the injuries that result from a crash. The recommended maximum speed limit in urban areas is 50 km/h, with lower limits around schools and certain other places, but only 46 countries have laws that meet best practice around speed. The second is *drink-driving*: it seems obvious that you shouldn't drink and drive, but only 35 countries have laws that meet best practice around drink-driving. The third, involving *seat belts*, fares better: 102 countries observe best legislative practice. The fourth involves the *wearing of helmets for motorcyclists*; 44 countries observe best practice. The fifth is *child restraints*: only 29 countries have laws that observe best practice. (Why do so many countries put their children at risk?)

Better laws would save lives. So, perhaps, would self-driving cars: this technology will presumably improve road safety, since computers don't get tired or angry or drunk, and the transport efficiencies it could deliver would have a significant impact on other areas of life too.

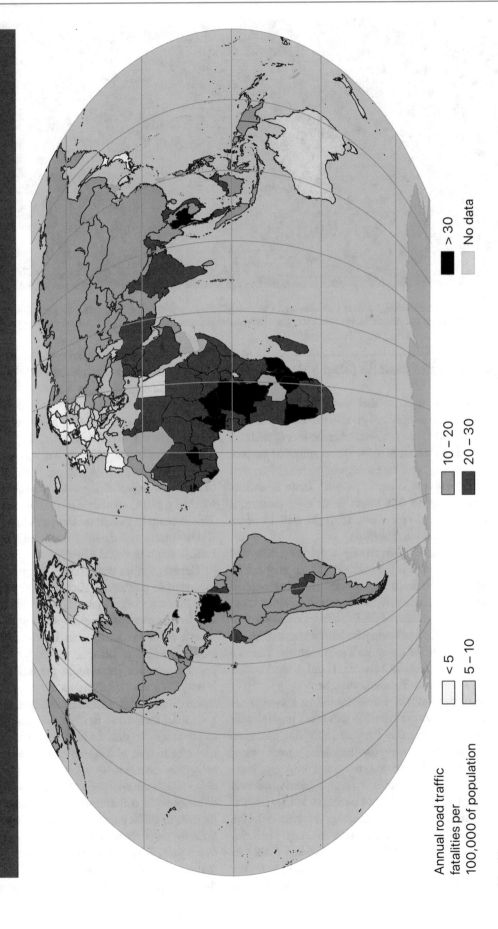

55 Road traffic deaths

Annual road traffic
fatalities per
100,000 of population

< 5
5 – 10
10 – 20
20 – 30
> 30
No data

Map 55

Fig. 42 A motorcyclist negotiating a road in a town about 15 km east of Liberia's capital city, Monrovia. Much of Liberia's road network remains unpaved, which is perhaps one reason for the country's poor record on road safety. (blk24ga, CC BY 3.0)

56   Incidence of Malaria (Map 56)

Picture, please, the creature that poses most danger to humans. What comes to mind? A great white shark, perhaps? A lion? A crocodile? Or do you remember the tens of millions of people who lost their lives in the blood-spattered wars of the twentieth century and conclude, quite reasonably, that people pose the greatest threat to people? There is another contender: the mosquito. The insect is far more dangerous than sharks, lions, and crocodiles. It might well be more deadly to humans than other humans.

Mosquitoes themselves are merely a nuisance. The real problem with mosquitoes is the ease with which they can be infected by *Plasmodium*, a group of single-celled microorganisms. In particular, the female *Anopheles* mosquito can carry the deadliest of the five *Plasmodium* species: *P. falciparum*. If an infected mosquito bites a human then the parasite can pass from the mosquito's saliva into the victim's blood. And then, some time between 1 and 4 weeks after infection, the symptoms of malaria start. See Fig. 43.

Malaria continues to take a dreadful toll on humanity. A patient with *P. falciparum* infection will often present with paroxysm, a cycle of sweating fever and shivering coldness. The initial flu-like symptoms—headache, joint pain, vomiting—are nasty, but other symptoms are worse: jaundice, seizures, and sometimes coma. Various common complications, such as infection of the brain, kidney failure, and severe anemia, can be life threatening. Malaria lacks the frightening mortality rate of such dread diseases as Ebola, Marburg, and the plague, but what it lacks in lethality in makes up for in prevalence. In 2018, according to the World Health Organisation (WHO), there were an estimated

228 million cases of malaria. Unlike a disease such as flu, which tends to move around the world in waves, occasionally causing widespread misery and then ebbing away before the next flare-up, malaria is a constant companion.

Year after year, malaria harrows humanity. In 2018, malaria caused 405,000 deaths. In 2017 malaria caused 435,000 deaths. In 2016 it was 451,000 deaths. In 2010, 607,000 deaths. And on and on. For comparison, according to currently accepted figures, over a period of 2 years the Covid-19 global pandemic has taken about 4.8 million lives. This is roughly five times the yearly rate of malaria deaths. But remember that the toll from malaria has been continuous, back through recorded history.

Genetic studies show the *P. falciparum* form of malaria increased in prevalence about 10,000 years ago, around the time humans started to develop the techniques of agriculture and live in permanent settlements. So it seems malaria has been a constant, deadly, dread companion. Indeed, the disease has killed or affected so many people it has put the greatest selective pressure on the human genome in recent history. We see its effects today. Some people suffer from sickle cell disorder, a genetic condition that distorts red blood cells from their usual flexible, concave shape into a curved, sickle shape. In distorted cells the hemoglobin molecule is less effective at picking up and releasing oxygen, and the cells themselves circulate less freely. If a person is unfortunate enough to carry two copies of the faulty gene then they develop sickle cell anaemia, a horrible affliction that causes patients to suffer attacks of pain from an early age and have their lives cut short. The disease kills about 55,000 people every year. Why does this genetic mutation not die out? The mutation remains because if a person has only *one* copy of the faulty gene then they don't suffer severe anaemia, while

56 Incidence of malaria

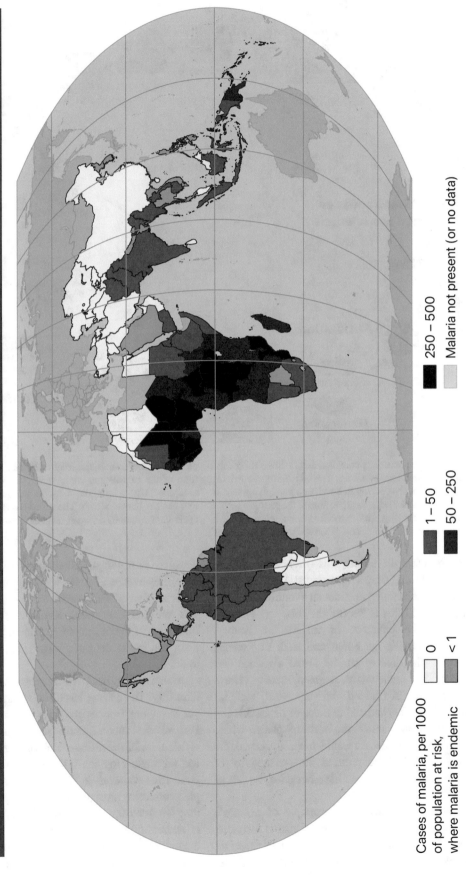

Cases of malaria, per 1000
of population at risk,
where malaria is endemic

0

< 1

1 – 50

50 – 250

250 – 500

Malaria not present (or no data)

Map 56

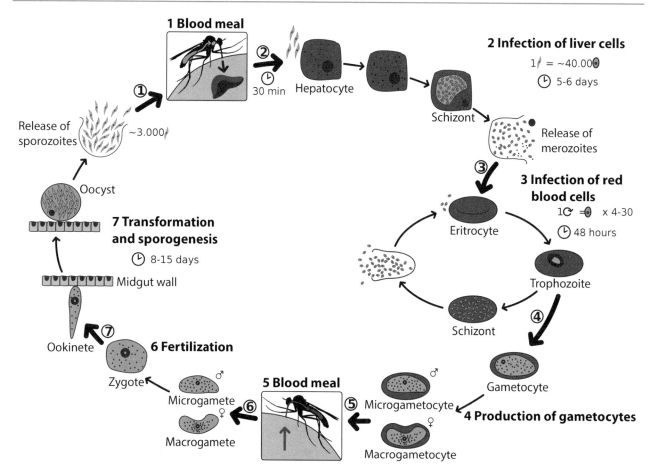

Fig. 43 The life cycle of malaria is complex, and involves two hosts: humans and female Anopheles mosquitoes. If an infected mosquito bites a human then the parasite can develop in the victim, multiplying first in the liver and then in red blood cells. Reproduction of the parasites within a red blood cell will eventually cause the cell to burst, releasing more parasites that invade other red cells. This is the stage where symptoms present. If another mosquito feeds on blood during this stage, a different form of the parasite can continue its life cycle inside the gut of the mosquito. After a fortnight or so, the mosquito can inject the parasite into another human victim. The parasite migrates to the liver, and the cycle starts again. (Own work, adapted from Bbkkk, CC BY 4.0)

the tendency to sickle the blood cells offers immunity to malaria. In other words, deaths from malaria put so much pressure on human selection that, from an evolutionary point of view, the sickle cell trait is 'worth' the cost.

We even see the influence of malaria on history. Archaeologists examining the bones and teeth of Romans who lived during the time of empire found evidence that malaria was prevalent in the warm, stagnant, marshy valleys of the Tiber and the Po. (Indeed, the very word 'malaria' comes from the Latin for 'bad air' because of its association with swamps and marshlands—the sorts of places where mosquitoes thrive.) The disease could have weakened the Roman army; it would have reduced the productivity of agricultural workers; it might well have hastened the fall of Rome itself.

We also have a more mundane example of malaria's influence on the modern world. Quinine, which is extracted from the bark of the cinchona tree, has a beneficial effect on malaria. People therefore began to drink quinine powder

dissolved in water: so-called tonic water. Quinine has a bitter taste, however, and tonic water containing a high enough concentration of the powder to be effective against malaria was not a pleasant drink. British army officers stationed in India, one of the hotspots for malaria, therefore added sugar, lime, and gin to the tonic water: thus was born the gin and tonic.

Malaria, then, has been a constant and consistent bane of humanity. The disease has claimed more lives than war. And yet it seldom reaches the pages of the newspapers. Why? Perhaps the uncomfortable reason is to be found in the distribution of the disease: malaria is endemic in the tropical and subtropical regions of the world, which also happen to be poor. Rich countries in temperate regions have either put in place effective controls or they enjoy conditions such that they are free from malaria. Mainland Australia, for example, is malaria-free. Although each year about 500 cases are diagnosed in Australia, these are invariably people who travelled to malaria-affected countries and failed to take anti-malarial drugs. If the disease were as prevalent in Europe and the US as it is in sub-Saharan Africa then surely we would see a 'war on malaria'. (We might yet: as the climate warms, mosquitoes will move.)

The December 2019 edition of the World Health Organisation's annual *World Malaria Report* states that, in 2018, 93% of the world's malaria cases occurred in the African region. Almost 85% of the global malaria burden was carried by 20 countries: 19 in sub-Saharan Africa, plus India. Just six countries accounted for more than half of all malaria cases: Nigeria (25%); Democratic Republic of the Congo (12%); Uganda (5%); Côte d'Ivoire (4%); Mozambique (4%); and Niger (4%).

The map shows the incidence of malaria per thousand of the population at risk. African countries again stand out. In Rwanda the incidence is 486 per thousand of population at risk. In Burkina Faso the incidence is 399 per thousand of population at risk. The Central African Republic and Mali both have incidences in excess of 386 per thousand of population at risk.

Rather than draw the map of malaria as a choropleth we could draw it as a cartogram, and represent a country's size by the total number of its malaria deaths. If the world map were so distorted then Nigeria would be the largest country on Earth (96,000 deaths), followed by the Democratic Republic of the Congo (45,000 deaths), and United Republic of Tanzania (20,000 deaths). Malaria is an African problem.

Humanity *is* pushing back. In 2017, after years of interventions, from the draining of swamps to the spraying of insecticides, China and El Salvador had no indigenous cases of the disease. In 2018, WHO certified Paraguay and Uzbekistan as malaria-free. And in 2019, science brought out another weapon in the war against malaria: genetic engineering. Biologists genetically modified particular gut genes in mosquitoes so that, after taking a blood meal, they released antimicrobial molecules that attack the *Plasmodium* parasite. The mosquitoes can successfully reproduce, passing their altered genes to the next generation. This approach is still in its early stages but the hope is, by releasing these genetically modified mosquitoes into the environment, the beneficial genes will spread among the population and the insects are less likely to transmit the malaria parasite.

Genetic modification is only one weapon in science's armoury: we now have a vaccine. Progress on vaccine development has been slow, in part because malaria primarily affects poor nations and in part because of the complexity of the parasite's lifecycle. In 2021, however, WHO recommended the use of RTS,S/AS01—the first vaccine shown in large-scale trials to be safe and effective against the disease—with children living in regions with moderate to high transmission of malaria. The vaccine itself will not solve the problem: treatment requires a schedule of four separate doses, which poses logistical difficulties for some sub-Saharan countries, and although the vaccine is effective the reduction in deadly severe malaria is only 30%. Many fully vaccinated children will still die of malaria. But this will be the first of many vaccines.

The global scientific response to Covid-19, and the heavy investment in RNA vaccine technology, is now being applied to other diseases. Immunologists at Yale University have developed a 'self-amplifying' RNA (or saRNA) vaccine for malaria. Human testing could start in 2024. Perhaps soon, for the first time in millennia, malaria will no longer be our partner.

57   Experience of Pain Yesterday (Map 57)

Gallup, the American-based analytics company, has its roots in the polling of public opinion. In 2005, it began a World Poll—a survey of over 150,000 citizens from more than 140 countries. Its *Global Emotions Report* is an attempt to measure intangible aspects of life, ingredients that traditional economic indicators fail to capture (here's looking at you, GDP; see Map 74). It aims to provide a "snapshot of people's positive and negative daily experiences". The Positive Experience Index asks five questions, such as "Did you feel rested yesterday?" and "Did you experience enjoyment yesterday?". The Negative Experience Index asks five questions, such as "Did you experience anger yesterday?" and "Did you experience stress yesterday?".

I am not a social scientist, so I cannot appreciate the survey's subtleties, but I find it hard to see how a country-by-country comparison of stress levels, for example, is illuminating. I experienced stress yesterday, but the feeling was not entirely negative: I was responding to a self-imposed deadline. The quality of the stress I experienced is different to that of a person living in a war zone. The Negative Experience Index, however, also asks: "Did you experience physical pain during a lot of yesterday?" Presumably our experience of pain *can* be compared.

Before the onset of the Covid-19 pandemic the response to this question was, to me, surprising. The global mean response in the year 2019 was that one in three people reported experiencing pain during a lot of the previous day. Is this not on the high side? There was variation, of course, as the map shows. Countries such as Belgium, Spain, and the USA were representative: here, 33% of the respondents felt physical pain during a lot of the previous day. The percentage in some other countries was lower: Vietnam (14%), Taiwan and Poland (15%), and Sweden (17%) reported much less pain. And then there was Chad.

Two in three citizens of Chad reported experiencing physical pain the previous day. Chad also topped the table in terms of the percentage of citizens who experienced sadness (54%), and it came second to Mozambique in terms of the percentage

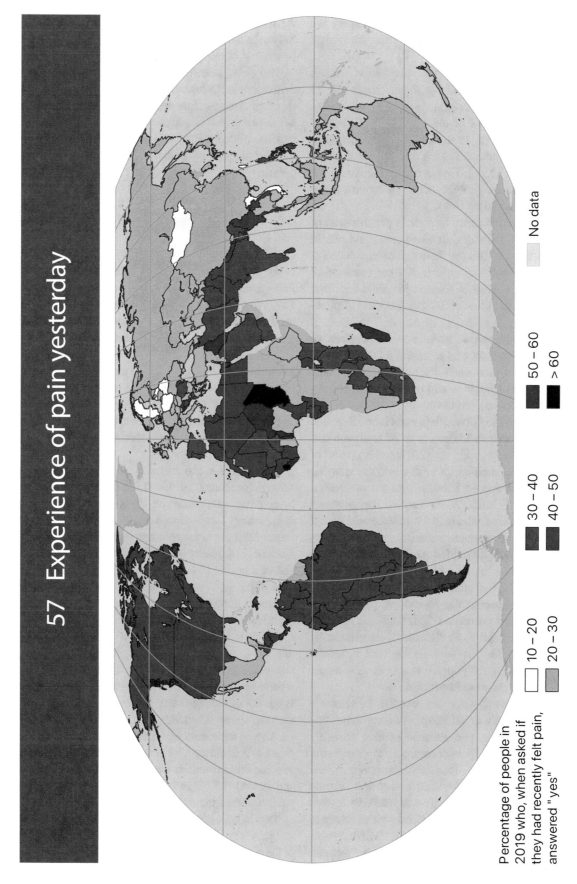

57 Experience of pain yesterday

Percentage of people in
2019 who, when asked if
they had recently felt pain,
answered "yes"

☐ 10 – 20	▨ 30 – 40	■ 50 – 60
▨ 20 – 30	■ 40 – 50	■ > 60

No data

Map 57

of citizens who experienced worry (61%). When Gallup published their Poll for that year, Chad was a country with problems: large numbers of Sudanese refugees had fled there to escape the war in Dafur; its young population had a life expectancy of just 52; and domestic violence against women was common. The country seems as if it would have been ill-prepared for the approaching pandemic.

I have not shown data from the 2021 *Report* because that covers the pandemic year 2020. Even if I had, we would not know the extent to which Chad's people experienced pain that year: Gallup were unable to collect data for Chad. One can hope the situation improved. Gallup, however, say that "in 2020, the world was a sadder, angrier, more worried and more stressed-out place than it has been at any time in the past 15 years" so one fears the worst.

Amongst the countries polled by Gallup in 2020, Iraq had the unwanted distinction of topping the table in terms of the experience of pain: 56% of Iraqi citizens had experienced a lot of pain in the 24 h prior to the question being asked of them. This was followed by Egypt (55%), Senegal (52%), Republic of the Congo (47%), and Lebanon (46%). At the more attractive end of the scale, Vietnam (16%) and Taiwan (15%) were once again placed highly; Hong Kong (14%), Mongolia (15%), and Israel (16%) complete the top five.

Although much of the Negative Experience Index is based on subjective feelings, the results do surely tell us *something*. In 2020, more than half of Iraqis experienced each of the five negative experiences and the country topped the lists not only of the experience of pain but also anger (51%) and sadness (50%). Perhaps Chad, had it been polled, would have scored worse. And one suspects that Afghanistan, also not polled, would have fared worse still.

---

58   How the World Dies (Map 58)

As with all such illustrations, you should interpret with care this map of the commonest causes of death. For one thing, the map refers to all deaths recorded in a country and makes no attempt to adjust for age. Clearly, the age profile of a population is relevant here: the most likely cause of death of a 9-month-old baby is not the same as of a 90-year-old adult. For another thing, it refers to a particular year—2019, to be precise—and such a map changes over time. For example, according to the World Health Organisation (WHO), in Syria and Iraq in 2016 the most likely cause of death was "collective violence and legal intervention". That would not always have been the case in the past and thankfully it was not the case in 2019. Another example: had someone drawn a similar map one century ago the appearance would have been entirely different because an influenza pandemic had just killed between 3% and 5% of Earth's population. As I write, around 5 million people have died of Covid-19—but the disease had yet to make an impact in the statistics for the year under discussion. As long as you approach the map with caution, however, it gives an idea of the sorts of ways people die in the early part of the twenty-first century.

According to WHO, the commonest cause of death worldwide is ischaemic heart disease, a somewhat old-fashioned term for what is now usually called coronary heart disease. This is the main killer in 67 countries across the world. Ischaemia is the medical term for an insufficient supply of blood to an organ. When an organ's blood supply is restricted then damage to tissue can occur because of a shortage of oxygen. And if the organ concerned is the heart—well, that's serious.

Coronary heart disease describes a situation in which the vessels that transport oxygen-rich blood to the heart muscle have narrowed—typically because of a build-up of fatty material, called atheroma, on the arterial walls. The condition is called atherosclerosis. If a piece of atheroma breaks off it can generate a blockage; if the blockage cuts off the oxygen supply to the heart muscle then the organ can be damaged and the person suffers a heart attack.

Coronary heart disease is the top killer of men and women of all races, and it is responsible for 16% of all the world's deaths.

Atherosclerosis can lead to another killer: stroke. If a blockage occurs in an artery carrying oxygen-rich blood to the brain, for example, then a stroke can occur. Stroke is the main cause of death in only 10 countries but, as it is responsible for 11% of total deaths worldwide, it is the second leading cause of death. I have no wish to sound gloomy but the chances are not unreasonable that, when you die, some form of cardiovascular disease will have been the cause.

In 60 countries, the leading cause of death is malignant neoplasms—essentially, cancer Although this map groups it as a single condition, in reality cancer consists of more than a hundred different diseases. The various diseases are characterised by the uncontrolled growth of abnormal cells, but cancer can arise in many organs and it behaves differently depending on its place of origin. Lung cancer, for example, has different characteristics to breast cancer.

If we were to separate out the different types of cancer then the map would look very different: no individual cancer is a leading cause of death, so those countries assigned a colour for malignant neoplasms would be assigned some other colour (in many cases the colour for ischaemic heart disease or stroke). Overall, though, cancer kills many people. Cancers of the trachea, bronchus, and lung kill most: worldwide, this is the sixth leading cause of death.

That cardiovascular disease and malignant neoplasms take so many lives is in part due to tremendous progress in preventing and treating communicable diseases. But communicable disease remains a problem in some countries.

58 How the world dies

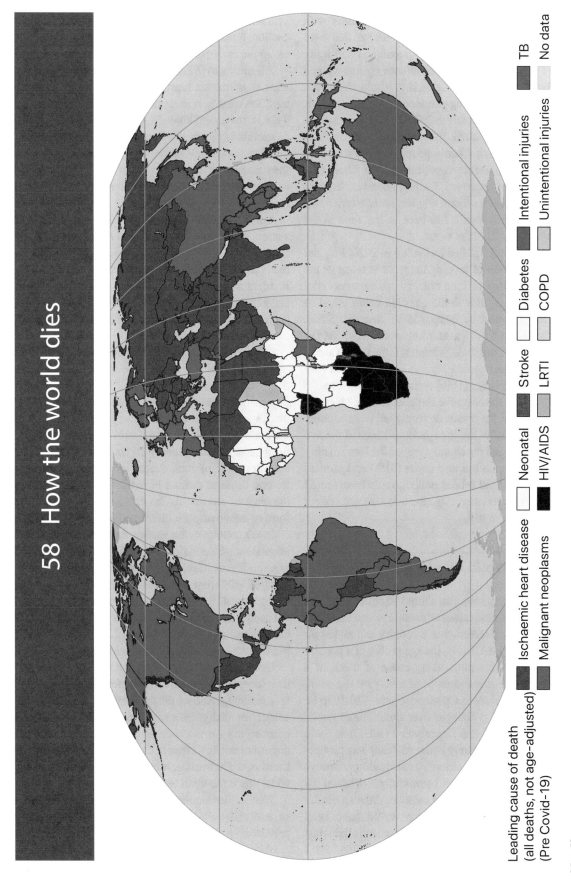

Leading cause of death
(all deaths, not age-adjusted)
(Pre Covid-19)

- Ischaemic heart disease
- Malignant neoplasms
- Neonatal
- HIV/AIDS
- Stroke
- LRTI
- Diabetes
- COPD
- Intentional injuries
- Unintentional injuries
- TB
- No data

Map 58

Although lower respiratory tract infections (LRTI) are the main cause of death in only six countries, globally they account for the highest number of deaths caused by communicable disease. LRTI represents the fourth leading cause of death across the globe.

Pneumonia is the commonest type of LRTI, and the commonest cause of death due to infection in Europe and the USA. In the developed world, however, patients can expect high-quality care: the severity of the disease will be assessed, the causative organism identified, and appropriate treatment delivered. In some African countries the death toll due to pneumonia is heavy—particularly amongst children.

Another significant killer is HIV/AIDS. This is the commonest form of death in 11 countries, all of them in Africa.

The human immunodeficiency virus, HIV, damages the cells of the immune system and leaves the host less able to fight off common infections. Acquired immune deficiency syndrome, AIDS, is the name given to describe the life-threatening infections that occur once HIV has damaged the immune system. In the developed world, patients who receive a timely diagnosis of HIV receive a cocktail of drugs that permit a long and healthy life. The virus can still be there in the body—there is as yet no cure—but the treatment ensures that AIDS-related illness do not develop. Thanks to this effective treatment and control, HIV/AIDS dropped from being the 8th leading cause of death in 2000 to the 19th in 2019. In countries that do not have those medical interventions, HIV/AIDS remains a mass killer.

Tuberculosis is a horrible affliction caused by airborne bacteria. It was once extremely common, and known as the 'romantic disease' because of the large number of artists it killed, but it is another communicable disease that now kills fewer people than before: the world saw a 30% reduction in the number of deaths between 2000 and 2019. In the Asian country of Timor-Leste, however, it remains the biggest killer.

The first weeks of life can be a dangerous time. In 23 countries, neonatal conditions—birth asphyxia and birth trauma, neonatal sepsis and infections, and preterm birth complications—was the biggest killer. Across the globe, it constituted the fifth leading cause of death. Annual deaths have reduced by almost 40% since 2000, however, so the trend is in the right direction.

Diabetes mellitus is the biggest killer in only three countries: Fiji, Kiribati, and Mauritius. Nevertheless, diabetes mellitus a cause of concern around the world because of its rising incidence (see also Map 53). If account is made not just of mortality, but also of the number of years of healthy life lost due to disability, then diabetes is a large and growing problem.

In Nepal, the biggest killer is chronic obstructive pulmonary disease (COPD). Nepal is the only country where COPD kills more people than anything else, but globally COPD is the third leading cause of death.

COPD is the name for a group of lung conditions. The conditions include emphysema (damage to the air sacs in the lungs) and chronic bronchitis (long-term inflammation of the airways), and they cause breathing difficulties. The Nepalese population has particular risk factors for COPD: smoking (more than 28% of Nepalese men smoke) and exposure to smoke from biomass fuel (two thirds of Nepalese households use biomass fuel to cook and to heat homes; see Fig. 44). But COPD takes a heavy toll everywhere.

In Guatemala the main cause of death is not disease but what the WHO class as 'unintentional injuries'. The term covers road injuries, poisonings, falls, fire, drowning, exposure to mechanical forces (a bang on the head?), natural disasters and other. In Guatemala, the major contributor to the class of 'unintentional injuries' was road injuries (see Map 55).

And in El Salvador, sadly, the main cause of death is what the WHO class as 'intentional injuries'. A combination of interpersonal violence (see Map 51) and self-harm (see Map 52) are responsible for more deaths in El Salvador than coronary heart disease, stroke, or cancer.

In looking at the leading causes of death in individual countries we have looked at many of the top ten leading causes of death worldwide. But some diseases, although not the biggest killer in any single country, have a large combined death toll. Diarrhoeal disease is the world's eighth leading cause of death. Diarrhoea drains the body of the water and salts we need for survival. For children under five the condition can be devastating. And yet the causes are so easily dealt with: most cases of diarrhoeal disease can be prevented simply be providing safe drinking water and adequate levels of hygiene and sanitation. Diarrhoea remains a problem in countries such as Burundi and Kenya, but the picture is improving.

The seventh leading cause of death globally is what the WHO class as 'Alzheimer disease and other dementias'. The risk of developing dementia, a syndrome associated with an inexorable decline in brain function, rises with age. As the population ages, the number of sufferers increases. In 2000 it caused just over half a million deaths and it ranked 20th in the list of leading causes of death; in 2019, the disease killed 1.6 million people.

If, as argued above, we consider cancer to be an umbrella term for different diseases rather than a single disease, then Alzheimer disease is already the leading cause of death in three countries—Finland, Netherlands, and the UK. In these three countries Alzheimer disease kills more people than coronary heart disease and stroke. In the UK, the disease affects 1 in 14 people over the age of 65, and 1 in 6 people over the age of 80.

The precise cause of Alzheimer disease is unknown. There is no cure.

Fig. 44 Children in Nepal are often exposed to smoke, a risk factor for COPD in later years. (Parth Sarathi Mahapatra, CC BY 3.0)

59   The Life Expectancy Gender Gap (Map 59)

Here is a remarkable fact. In every country on Earth, a newborn girl can expect to live longer than a newborn boy.

The size of the gender gap varies between countries. In Bhutan, at the start of 2019, a female baby could expect on average an extra 7 months of life compared to a male baby born at the same time. In Syria, the difference was 12 years. But why does the gap exist at all? Most societies have in-built structural advantages for males, so why should women live longer?

The answer to that question is unclear. Biology certainly plays a role. One possibility is that the male hormone testosterone provides a short-term benefit by increasing body strength in early years at the long-term cost of increasing the chance of developing heart disease and cancer in later years. Some tentative evidence exists to support this idea: eunuchs castrated before the age of 15 have been observed to live longer than average. But presumably behaviour and environment also play a role. The relative contribution of biology, behaviour, and environment to increased female life expectancy is uncertain.

If behavioural and environmental factors do indeed play a role, what might these factors be? The specific example of the UK might provide some general insights.

First, in the twentieth century, two world wars took a heavy death toll on men. Over 800,000 UK citizens died in military action in World War I, and the bulk of those deaths were of young men. In World War II, about 380,000 UK citizens died in action. Deaths from conflict have thankfully decreased in the UK, but in Syria, the country with the largest gender difference in life expectancy, a conflict has burned for 10 years. A peaceful uprising in March 2011 descended into a civil war that has so far killed almost 400,000 people. The combatants are more likely to be men than women.

Second, patterns of cigarette smoking—and thus various forms of cancer—differ between the sexes. People are now smoking less in the UK, but men have always smoked more than women. Across the world, 35% of men smoke compared to 6% of women.

59   The life expectancy gender gap

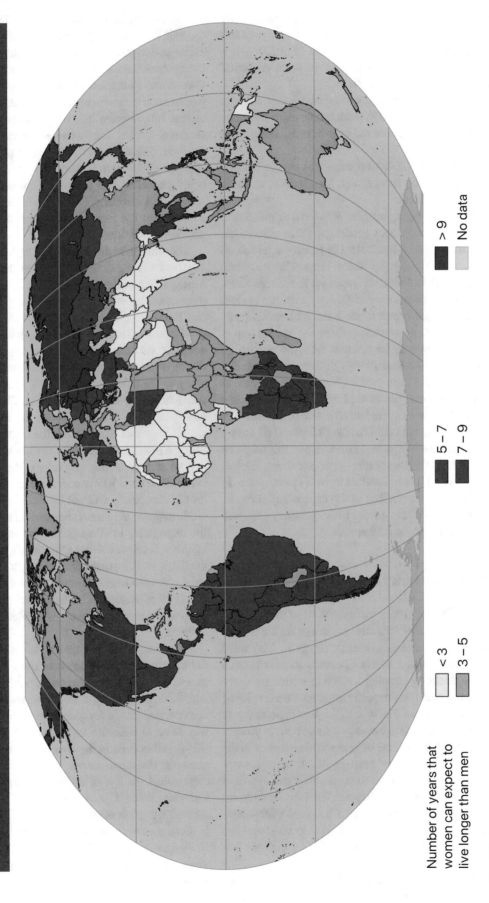

Number of years that
women can expect to
live longer than men

< 3

3 – 5

5 – 7

7 – 9

> 9

No data

Map 59

Third, in the UK men drink more than women. Heavy alcohol consumption (see Map 54) is dangerous. It causes disease, it makes suicide more likely (see Map 52), and it increases the chance of fatal accidents. In some countries, men drink *much* more than women.

Fourth, males are more likely than females to work in dangerous industries. In the UK, work is now more sedentary for both sexes. But this is not the case in many developing nations.

Fifth, men are more likely than women to engage in high-risk behaviour. Men are more likely, for example, to die in high-speed road accidents.

In the UK, the changing patterns of behaviour mentioned above mean the life expectancy gender gap for its citizens has narrowed in recent times: in a 2019 UN report the gap stood at 3.6 years. In 1980, the gap was 6 years. The long-term impact of Covid-19 on the gender gap is yet to be determined, but over the coming decades we might expect the gap to close further. And we might expect to see a similar trend in many developed countries. In countries where the gender gap is currently widest, however, the future is harder to predict.

If we ignore Syria, whose 12-year difference in male and female lifespans arises from conflict, the biggest gaps are in countries that were once part of the USSR. Lithuania (with a gender gap of 10.6 years), Russia (10.3 years), Belarus (9.9 years), Latvia (9.8 years), and Ukraine (9.7 years) all have a high mortality rate among younger men. In part this might be due to drink: alcohol-related deaths—from accidents, suicide, and disease—are all more prevalent in men than women. In these countries an improvement in life expectancy for men, and thus a reduction in the gender life expectancy gap, will come when attitudes to vodka improve.

60   Life Expectancy at Birth (Map 60)

According to Psalm 90:10, in the translation of the Bible by Myles Coverdale in 1539, "the days of our age are three score years and ten". A score is a scratch or notch. The word became associated with a number because shepherds once counted large flocks by scraping a notch in a stick for every 20 sheep. To use much less majestic language than employed by Coverdale, then, the Psalm tells us life expectancy at birth—the mean lifespan we can expect a newborn to possess if mortality patterns at the time of birth remain constant in the future—is 70 years. Babies born in 2019, in a country coloured red on this map, were fortunate: they can expect at least an extra 10 years of life beyond 70. Threescore years and ten and ten. (As we saw in Map 59, women everywhere can expect to live longer than men. This map illustrates the combined life expectancy in a country, rather than the life expectancy for one sex.)

Life expectancy at birth is a statistical construct. Consider a pair of twins, one of whom lives to be 70 while the other dies before reaching a first birthday. The life expectancy of the pair is 35, which provides little illumination. But for an entire society the number tells us something about general living conditions. In the Bronze and Iron Ages, and in ancient Greece and Rome, life expectancy at birth was about 26. A person back then who celebrated a 26th birthday had a decent chance of surviving into old age, but high levels of infant mortality pulled the average down. By early Victorian England, life expectancy at birth had improved, but only slightly: a baby born in the year Victoria ascended the throne could expect to live to see the Queen's 40th Jubilee. It is advances in medicine, and in public health in particular, that have driven the increase in life expectancy those of us in privileged countries enjoy.

The three countries with the highest life expectancy in 2019 were Hong Kong (84.8 years), Japan (84.6 years), and Macau (84.2 years). This is about *30 years longer* than the three countries with the lowest life expectancy: Central African Republic (53 years), Lesotho (53.9 years), and Chad (54.1 years). The gap in life expectancy between African countries and the rest of the world is stark. Citizens of the disputed state of Western Sahara have the closest life expectancy to the Biblical figure of 70 years; of the 61 countries with a life expectancy less than that of Western Sahara, 45 are in Africa. The 27 countries with the lowest life expectancy are all in Africa. Addressing the causes of this disparity is surely a moral imperative for the world.

The numbers in this map refer to the period just before the onset of the Covid-19 pandemic. The effects of the pandemic on life expectancy will not be clear for some time, but early work suggests the disease has caused the biggest decrease in life expectancy in Western Europe since World War II; the USA has been even worse affected, with male life expectancy declining by 2.2 years compared to 2019. The impact of the pandemic on African countries is not yet quantified, but since Covid-19 appears to be widening existing health inequalities everywhere we can assume that Africa will have been badly affected.

It is worth noting that the *maximum* human lifespan seems not to have increased much over time. In rich countries more people are pushing the longevity limits: in Okinawa, Japan and Sardinia, Italy, for example, you can find clusters of centenarians. But a few people born in the seventeenth century lived to celebrate their 100th birthday; a long lifespan was possible even in the absence of modern medicine. And none of the centenarians of Okinawa or Sardinia have approached the age of the oldest person in history, Jeanne Calment. When Calment was born in Arles in 1875, the average life expectancy was 43; she died in 1997, aged 122.

On a personal note, as I get older I find I'm less interested in *lifespan* and more interested in *healthspan*. The number of years spent in good health is more important than the final score.

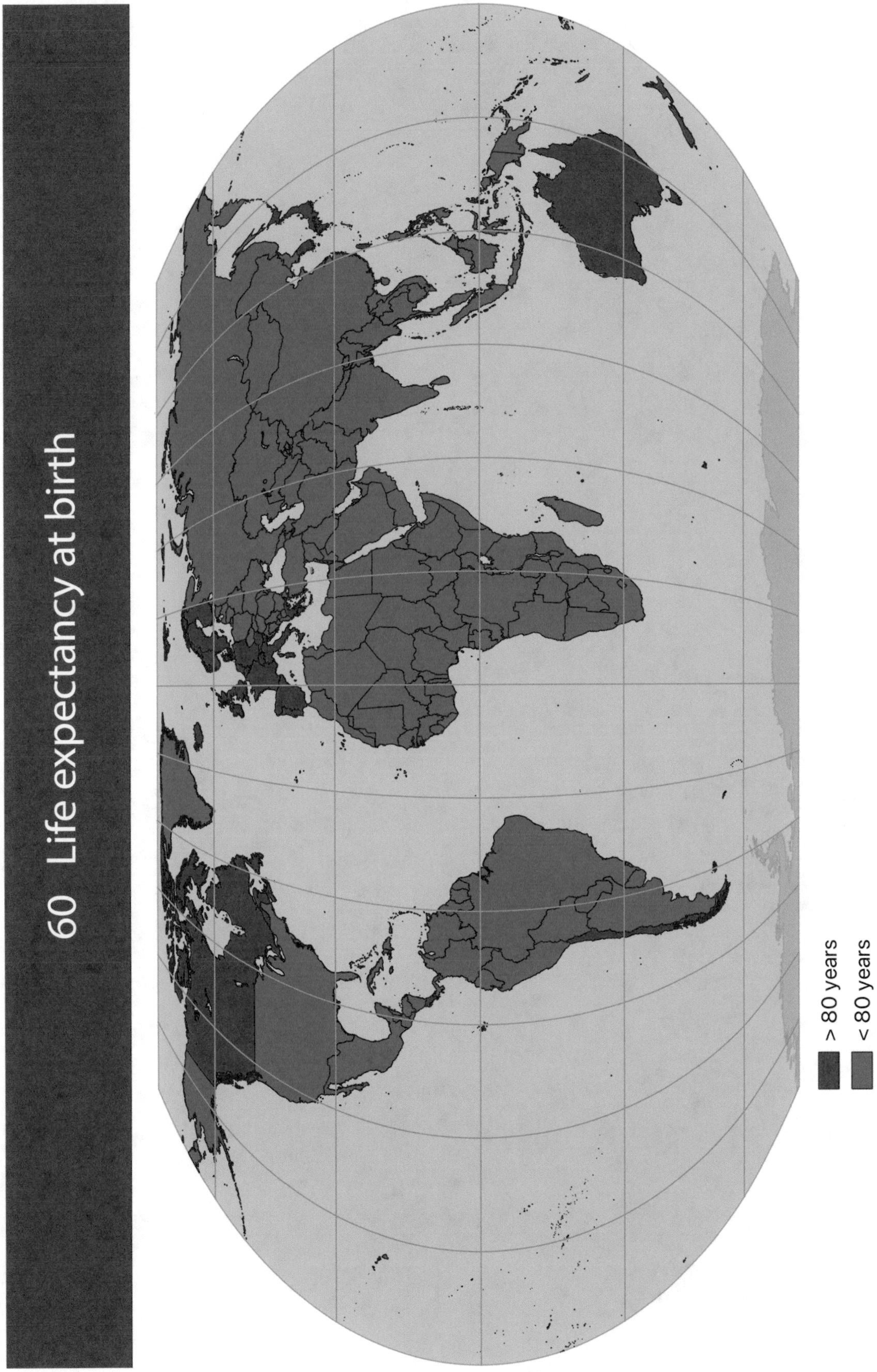

60 Life expectancy at birth

■ > 80 years
■ < 80 years

Map 60

The World of Sport and Leisure

Once life's necessities have been provided for—once food, water, and shelter have been obtained—people seek out ways to pass the time allotted to them. As we saw in the previous chapter, that time might be 80 years or more for those lucky enough to live in certain developed countries; it is decades shorter for the citizens of many African countries. However long or short the lifespan, though, we want to be entertained.

Sport, both participating in and spectating, is popular around the planet. So the maps in this chapter take in, for example, football, the global game, as well as the Olympic and Paralympic Games.

The maps illustrate aspects of three other near-universal forms of entertainment: theatre, literature, and movies.

And, since the digital world has become increasingly important in our lives, the maps examine various ways in which we use computers to pass the time.

61   Football World Cup (Map 61)

The world's most popular sport is football (or, to give it its formal name, association football—the first word of which, in contraction, supplied us with its other name: soccer). People play the 'beautiful game', as Pelé called it, in every nation. Indeed FIFA, football's international governing body, has more member states than the UN. The FIFA World Cup is the world's most watched sports competition: 3.572 billion viewers—more than half the world's population above the age of 4—watched part of the 2018 World Cup.

Since the inaugural event in 1930 the World Cup has been held every 4 years, with the exception of 1942 and 1946, when more pressing events were on people's minds. (In 1909 Sir Thomas Lipton, of tea fame, organised a competition in Turin in which individual clubs, rather than national teams, competed on behalf of their country. The Lipton trophy has been called the first football World Cup. I am inordinately proud that an amateur team from West Auckland, a mining village not far from the team I support, Middlesbrough, won the tournament and successfully defended it 2 years later. For fans around the globe, however, the term 'World Cup' means the FIFA World Cup.) So—which countries play the most successful brand of football, as measured by success in this competition?

As the map makes clear, the World Cup winning nations cluster in two regions: South America and Western Europe. Brazil, the only country to have qualified for every World Cup competition, has won five times (and hosted twice; see Fig. 45). Germany and Italy have both won four times; Argentina, France, and Uruguay have all won twice; and England and Spain are one-time winners. The five nations that have reached the final but never gone on to win it all come from Europe. And nations whose footballing pinnacle is the third-place play off (a pointless game if ever there was one) all come from Europe and the Americas.

Asia is clearly not a hotbed of football. More surprisingly, perhaps, teams from Africa have not performed well. In 1977, Pelé predicted that an African team would win the World Cup before 2000. The majority of African countries have still to qualify for a World Cup tournament. Of those that have qualified, the furthest progress has been the quarter final stage. Perhaps the World Cup in Qatar in 2022 will be the one for Africa?

© The Author(s), under exclusive license to Springer Nature Switzerland AG 2023
S. Webb, *Around the World in 80 Ways*, https://doi.org/10.1007/978-3-031-02440-5_8

61 Football World Cup

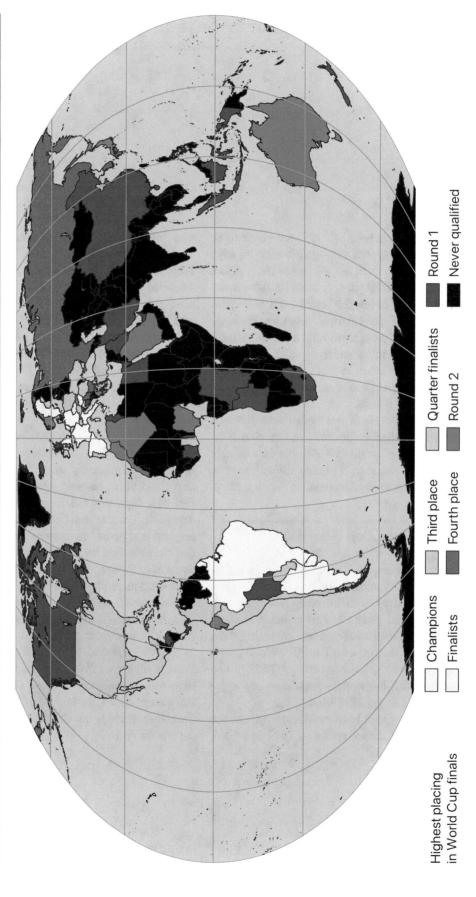

Highest placing
in World Cup finals

- Champions
- Finalists
- Third place
- Fourth place
- Quarter finalists
- Round 2
- Round 1
- Never qualified

Map 61

Fig. 45 The Maracanã Stadium in Rio de Janeiro, Brazil, was built for the 1950 World Cup. The attendance here at the 1950 final, in which Brazil lost 2–1 to Uruguay, remains a record attendance for a football match (see Map 40). The Maracanã also played host to the 2014 World Cup final, in which Germany beat Argentina 1–0. (Daniel Basil, CC BY 3.0 BR)

62 Success at the Summer Olympics (Map 62)

As a sports-mad child, long before the term 'binge-watching' had been invented, I gorged on television coverage of the Summer Olympic Games, the world's most prestigious general sporting event. I most enjoyed the athletics events but I happily watched sports whose rules I didn't understand and I even watched activities I didn't particularly like. Over the years, the Olympics has provided me with some of my most vivid memories of televised sport: astonishment at Usain Bolt's performances; shock at Ben Johnson's victory in the 100 m and his subsequent disqualification for anabolic steroid use; absorption in the ongoing personality clash between Seb Coe and Steve Ovett as they battled for middle-distance supremacy. And, every 4 years, there was the joy of the medals table—watching it update being almost as much fun as studying a football league table after the Saturday results came in.

But even as a kid I had a problem with the Olympics medal table.

If the USA was not at the top of the table then the USSR would be there. That in itself was not the source of my problem. Rather, what offended me—I was a strange child—was the attack on my sense of fairness. A country such as the USA would *of course* be high in the medal table: a country with a population measured in the hundreds of millions had a clear advantage over a country with a population measured in the hundreds of thousands.

Many people, I now know, were equally offended by the medal table, and by how the 'big' countries—USA, Russia, China—dominated. Every 4 years, therefore, around the time of the Summer Olympics, many newspapers have decided to provide an adjusted 'all-time' table of medals awarded (either gold medals or total medals). After the 2016 Games of Rio de Janeiro, Finland was the top nation in terms of total medals per capita, a legacy of earlier success in athletics and wrestling. By 2016, Finns had been awarded 302 medals of all types and 101 gold medals in Olympic Games dating back to 1908. With a current population of just under 5.5 million, this resulted in a ratio of 55 medals per million of population and 18.39 golds per million of population. (Paavo Nurmi, the 'Flying Finn', on his own won nine gold medals and three silver; another runner, Lasse Virén won four golds.) The Finns were less successful in the Tokyo 2020 Games, which the Covid-19 pandemic caused to be postponed to 2021, winning just two bronze medals. This allowed tiny Bahamas to overtake them in this adjusted league table. A Bahamian population of less than 290,000 people has produced 8 gold medals and 16 Olympic medals in total, which means they have won roughly 56 medals per million people. Countries such as the USA, Russia, China—and major European sporting countries such as the UK, Germany, France—do not appear in this adjusted top 10 list.

Although it is satisfying to see smaller countries ahead of sporting powerhouses such as the USA and Russa, tables

62 Success at the Summer Olympics

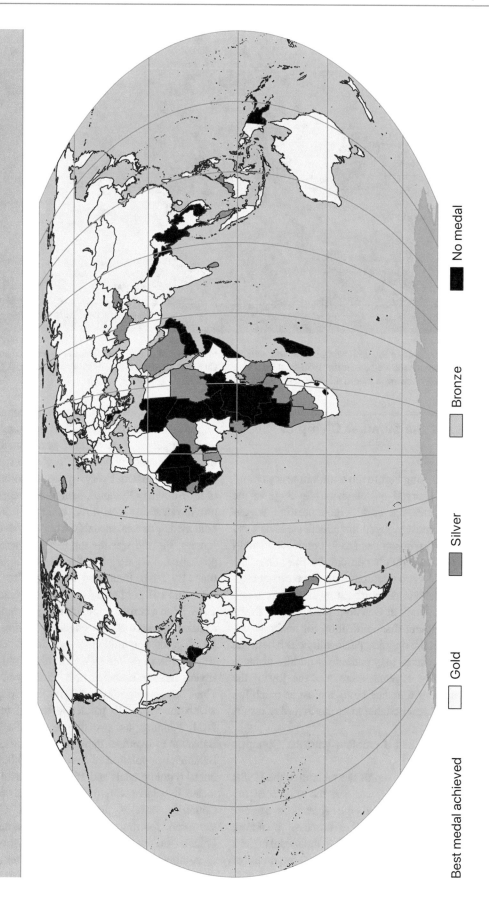

Best medal achieved

Gold · Silver · Bronze · No medal

Map 62

Fig. 46 The Panathenaic Stadium, in Athens, is on the site of a sporting arena built around 330 BCE. About five centuries later it was rebuilt in marble by the Romans and had a capacity of 50,000 seats. After about 1500 years of disuse the stadium was excavated, refurbished, and used for the opening and closing ceremonies of the first modern Olympics in 1896. Every 4 years, in a ceremony here at the Panathenaic Stadium, the Olympic flame is handed over to the host nation of the next Games. (George E. Koronaios, CC BY-SA 4.0)

based on per-capita medals are potentially as misleading as the raw medal table. They use a country's current population to calculate the statistic, but many countries won many of their medals at early Olympic Games when their population was much smaller. And population growth has not been uniform across all countries over time.

Another problem with these adjusted tables is that the number of medals available to be won has not been the same at each Olympic Games, nor has the number of nations competing. In the first Games of the modern era, the Athens Olympics of 1896 (see Fig. 46), only ten nations competed (along with a mixed team with individuals from those ten different nations; the gold medal in doubles tennis, for example, was won by a pair from Great Britain and Germany). In the 2020 Games, by comparison, athletes from 88 different National Olympic Committees appeared on the medal podium.

A further problem with adjusted tables is that countries sometimes miss a Games. Only five countries have competed at every Summer Olympics: Australia, France, Greece, Switzerland, and the UK. (The UK competes as Great Britain. This naming omits not only Northern Ireland, which is part of the UK and whose athletes compete for 'Team GB', but also other places that are not in the UK but whose athletes are eligible to compete for 'Team GB', including the Channel Islands, Isle of Man, and various other UK Overseas Territories. It's complicated.) And of course, over the span of more than a century of Olympic Games, the borders of a country can change. The USSR dominated the Olympic Games of my youth; today that union of countries no longer exists.

It is also worth mentioning that the sort of event for which medals are given has changed over time. In the 1900 Games in Paris, for example, live pigeon shooting was an event—as was croquet, obstacle swimming, and the long jump for horses. In the first five Olympic Games rope climbing was an event (in the 1904 Games in St Louis, George Eyser climbed up 25 ft of rope from a standing start, all the more impressive because he had a wooden leg). In Paris, St Louis, London (1908), and Stockholm (1912) the high jump was from a standing start. The changing nature of sports does not necessarily in itself invalidate a per capita calculation, but it is yet one more indication that comparisons across time are not easy.

So a 'fair' medal table would take account of a number of factors: the population of each country at the time of each of the Olympics; the number of nations competing at each of the Olympics and the total population represented by those nations; the number of medals available; and more beside. If you do that then, depending on the details of your model, it turns out that Finland remains an overachiever, if perhaps slightly overshadowed by New Zealand. But rather than try to create such an all-time per-capita map, I have instead opted simply to present a map showing the best medal achieved by each country in any Summer Olympics. The map takes into account results at the postponed Tokyo 2020 Games where, for example, Burkina Faso, San Marino, and Turkmenistan achieved their first Olympic medal.

Most countries have produced at least one gold-winning performance at the Summer Olympics. For countries such as Saudi Arabia, Iceland, and Ghana, silver is the highest place achieved on the podium. For a smaller number of countries, countries such as Afghanistan, Djibouti, and Iraq, the best performance is bronze.

Some countries, though, have never finished in the top three of any event. Some of those countries have tiny populations and so the lack of an Olympian is at first glance unsurprising. Antigua and Barbuda, for example, have an estimated population of only 96,000 and so one could argue that, statistically, the islands are unlikely to produce a great athlete. Having said that, Antigua and Barbuda have produced some of the greatest cricketers in history: one of the finest batsmen, Viv Richards, and two of the finest fast bowlers—Andy Roberts and Curtley Ambrose. Grenada has a similar population to Antigua and Barbuda and has produced a gold medallist: Kirani James, winner of the 400 m at the London 2012 Games. Some countries even smaller than Antigua and Barbuda have produced Olympians: Bermuda, for example, has a population of only 61,000 and yet has produced a bronze medal winning boxer (Clarence Hill at Montreal 1976) and a gold medal winning triathlete (Flora Duffy at Tokyo 2020). And San Marino, which has a population of only 34,000, won three medals at Tokyo 2020: bronze in the women's trap shooting, bronze in the men's freestyle 86 kg class wrestling, and silver in mixed trap team shooting. So small countries can from time to time compete at the highest level in the Olympics; the surprise lies in those countries with large populations that have never produced an Olympian.

Bangladesh has competed in every Games since Los Angeles 1984, so ten times in all. With a population of about 165 million people, it is the country with largest population never to have won a medal. The Democratic Republic of the Congo is the country with the largest land area never to have won a medal (and with a population of 90 million people it is, after Bangladesh, the second most populous country to be without an Olympic medal). In South America,

Bolivia is the only country to be entirely without success. Bolivia, which has a population of about 12 million people, first competed at Berlin 1936 and has participated in every Olympic Games since Tokyo 1964 (with the exception of Moscow 1980, which many countries chose to boycott because of the Soviet invasion of Afghanistan). Bangladesh, Bolivia, and DR Congo all have larger populations than those two sporting overachievers of Finland (5.5 million) and New Zealand (5 million).

For some countries, of course, the Winter Olympics hold even more importance than the Summer. Identifying an all-time list of Winter Olympic medals, however, seems particularly unfair, given the inherent advantage of countries that happen to be situated at certain latitudes! For what it is worth, the nation that has won the most gold medals, the most silver medals, and the most bronze medals in the Winter Olympics is Norway. (A further complication with all-time medal lists is that, as mentioned above, countries themselves change over time. Germany has competed as a single entity, but it has also competed separately as East Germany and West Germany. Taken together, Germany has been even more successful than Norway at winter sports.)

63   Success at the Summer Paralympics (Map 63)

The 1948 Olympic Games opened in London on 29 July. The same day Ludwig Guttmann, a German neurologist who had fled to England from Nazi Germany just before the outbreak of war, opened a sporting competition at Stoke Mandeville hospital. Guttmann's games were for servicemen and women with spinal cord injuries; 16 wheelchair athletes took part in an archery competition. As the director of a unit devoted to treating spinal injuries, Guttmann was more interested in the therapeutic potential of the games than the medal table. He believed sporting activity could help his patients rebuild self-respect and physical strength. His Stoke Mandeville games grew, and began to attract international athletes. The Olympic Committee took notice and in 1960, 1 week after the ending of the Rome Olympics, the first Paralympic Games began in the eternal city. That first Games saw 400 athletes from 23 countries compete in eight sports; due to the Games' origins, the only disability represented there was spinal cord injury. Since then, the Summer Paralympics has taken place every 4 years and the range of sports and disability classifications has increased with each Games.

Since the 1960 Rome Games up to the 2020 Tokyo Games the three most successful nations, in terms of total medals won, have been: United States (2283 medals); United Kingdom (1914 medals); and China (1237 medals). One might have predicted this: the UK has a long history of participation; the US and China have large populations; and all three

63   Success at the Summer Paralympics

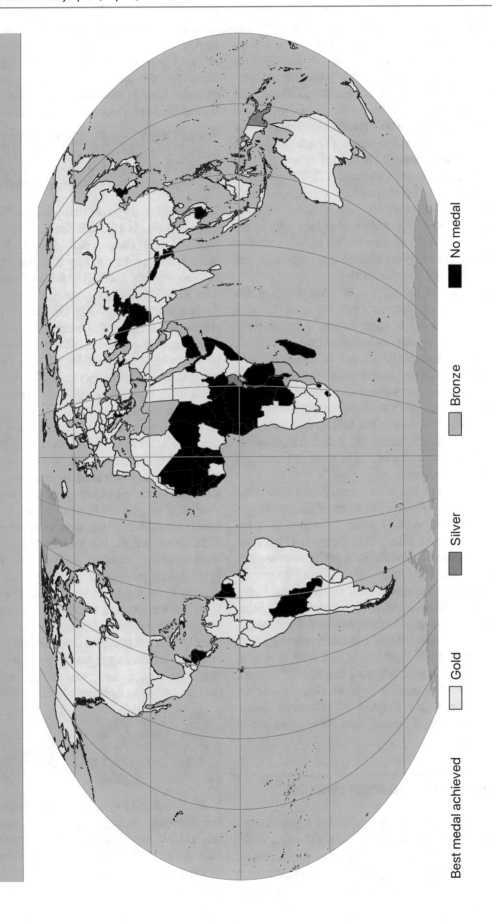

Best medal achieved   Gold   Silver   Bronze   No medal

Map 63

countries have the means to purchase the adaptive devices and technologies necessary for elite athletes to compete at the highest level. So, as with the Olympics, it is difficult to compare nation with nation. Indeed, producing a 'fair' medal table for the Paralympics is even more difficult than it is for the Olympics. In addition to the factors mentioned with Map 62, the range of Paralympic events is a confounding factor. The 100 m sprint, for example, offers medals in the classifications: T11–T13 (blind and partially sighted); T20 (intellectually disabled); T32–T34 (cerebral palsy, wheelchair); T35–T38 (cerebral palsy, ambulant); T40–46 (ambulant athletes with amputations or other disabilities); and T51–T58 (wheelchair athletes with spinal cord injuries or amputations). So as in the previous section, here I simply present a map showing the best medal achieved by each country in any Summer Paralympics.

Maps 62 and 63 are broadly similar. Most countries that have won a gold at the Summer Olympics have also won gold, or at least a medal of some colour, at the Summer Paralympics. The few exceptions include Armenia, Burundi, Cameroon, North Korea, Suriname, and Tajikistan. It is perhaps unsurprising that most countries without a medal at the Summer Olympics are also without a medal at the Summer Paralympics. This is the case for a number of African nations; for Bolivia in South America; for Belize, Honduras, and Nicaragua, in Central America; for Bangladesh, Bhutan, Cambodia, Nepal, and Turkmenistan in Asia; and a number of small nations and states.

I suspect some sports fans still view the Paralympics as the smaller cousin of the Olympics. In terms of the numbers of athletes that view is certainly valid: 2.5 times as many athletes competed in the Tokyo 2020 Olympics as the Paralympics. But in some ways the Paralympic Games now outdo the traditional Olympic Games. The Tokyo Paralympics offered 539 medal events compared to the 339 medal events on offer in the Olympics. And in several races, paralympic athletes record similar times to their able-bodied colleagues. In the swimming pool, for example, the current world record for the 100 m butterfly is 49.45 s, held by Caeleb Dressel of the USA. Dressel won gold in Tokyo. Ihar Boki of Belarus, who swims in the S13 visually impaired class, is only 4 s slower. On the track, for distances of 800 m or longer, wheelchair athletes are faster than their able-bodied colleagues.

64   Chess Playing Strength (Map 64)

In 1972 the world's sporting attention, before it turned to the Games of the XX Olympiad, fell—and this still seems rather bizarre—on chess. The eccentric American genius Bobby Fischer challenged the defending world champion, the dour, efficient Boris Spassky of the Soviet Union. Newspaper journalists and television crews spent the summer in Reykjavík, where the game took place, reporting how these two men pushed wood over a black-and-white board. The drama lay not just in the opposing personalities of the contestants but in what they represented: this was the Cold War playing out with chess pieces rather than bombs.

> Chess is a war over the board.
> The object is to crush the opponent's mind.—Bobby Fischer

The mercurial Fischer triumphed, but when he refused to defend his title in 1975 the Soviet Union resumed its domination of world chess, a position it had held throughout the quarter of a century before his victory. Anatoly Karpov became the new champion, followed by Garry Kasparov, then Vladimir Kramnik. Although neither nation has produced a world champion in the past 15 years, their overall chess-playing strength continues to this day. The International Chess Federation FIDE ('chess' is 'échecs' in French, hence the acronym) publishes a ratings list of chess players. A country's colour on this map refers to the average FIDE rating of its top-10 players as of October 2021. Russia and America stand out.

The ratings system is not perfect. Many experts, for example, rate Fischer at his peak as the strongest player of all time yet he appears only in 19th place on the all-time list. Ratings, furthermore, vary over time. Nevertheless, the average of the top-10 players in each country gives a rough indication of the strength of the game in those countries. As of October 2021, Russia tops the rankings with an average top-10 rating of 2731 points. Russia has held the top spot for years: it is undeniably the strongest chess-playing nation. The USA (2712 points average) is the only other country with the average of its top-10 players above 2700. Since 2015, the USA has not dropped below fourth place in the rankings. China (2699 points) is another strong chess-playing country: since 2015 it has occupied either second or third place in the rankings. To hammer home the point about Russia's chess-playing strength, note that it has most grandmasters (241, compared to America's 98 and China's 48).

> In chess the rules are fixed and the outcome is unpredictable. In Putin's Russia the rules are unpredictable and the outcome is fixed.—Garry Kasparov

If we omit Russia and the USA, 14 nations have a top-10 average rating of 2600 or above. These nations include England, but not Scotland or Wales because, as with FIFA, FIDE treats the UK countries as distinct entities. A further 28 countries have a top-10 average rating between 2500 and 2600. This tier of countries includes Norway, which is home

64 Chess playing strength

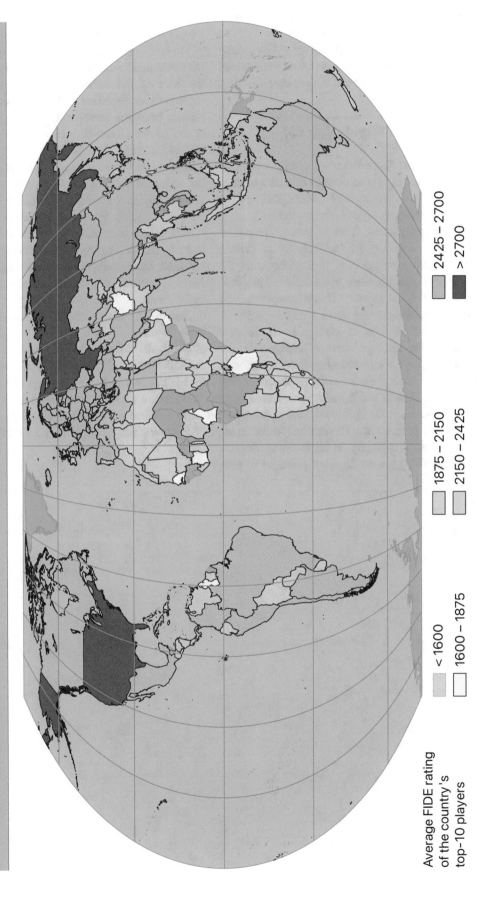

Average FIDE rating
of the country's
top–10 players

< 1600

1600 – 1875

1875 – 2150

2150 – 2425

2425 – 2700

> 2700

Map 64

to the current world champion and highest-ranked player: Magnus Carlsen (whose current rating is 2855).

African countries fare poorly. Only Egypt (2470 points) has had a strong chess-playing pedigree in recent years. Since 2015, Egypt's world ranking has been in the range 45–51.

65 Shakespeare's World (Map 65)

I sometimes wonder what Shakespeare—a man in possession of unbounded curiosity but only an average sixteenth century education—thought about the world as a whole. Did he have any conception our planet as we now understand it?

Shakespeare certainly knew Earth is round. He makes reference to Earth's sphericity in *The Comedy of Errors*, when he has Dromio compare Nell to a globe. And for Europeans of his time, Earth's shape was a matter of practical concern. The Bard was just 16 when Drake became the first Englishman to circumnavigate the world. Drake sailing back into Plymouth must have been as exciting to the teenage William as the Moon landings were to teenagers in the 1960s. He wrote his plays in a period when European powers promoted exploration as a key element of government policy. The Dutch, English, French, and Spanish were establishing colonies in North America; the Portuguese and Spanish had settlements in South America; the English had 'discovered' Muscovy and set up a company to trade there; English explorers were searching for northern sea passages to gain quicker access to the riches of Asia. The name of Shakespeare's theatre, 'The Globe', comes as no surprise.

On occasion Shakespeare made reference to distant lands: tales of Russia were in fashion; mention of India suggested the ends of the Earth; descriptions of Bermuda perhaps influenced his depiction of the (invented) island in *The Tempest*. In the main, however, he set his plays within the limits of the known Greco–Roman world. The action of 14 plays took place in historic Britain—a mythic Britain in the case of King Lear and Cymbeline. (If you believe Shakespeare was the author of *Edward III*, as several scholars now do, then you can bump the number of British plays up to 15.) Italy was the prime setting for ten plays; we don't know if he ever visited. He set three plays mainly in Greece; two in Turkey; and one each in Austria, Denmark (Fig. 47), Egypt, France, Lebanon, Spain, and Albania (assuming the Illyria of *Twelfth Night* is present-day Albania; several places in the Balkans could correspond to coastal Illyria.)

For those keeping count, I have omitted *The Tempest* (with its invented setting), *As You Like It* (also invented, although the Forest of Arden might be intended to evoke the Ardennes), and *Edward III* (since authorship is disputed). From the remaining 36 plays, though, we can see Shakespeare chose to constrain his imagination to settings in the classical world. Why? Was it the commercial decision of a jobbing playwright? Or did he lack a true feel for the 'globe'?

65   Shakespeare's world

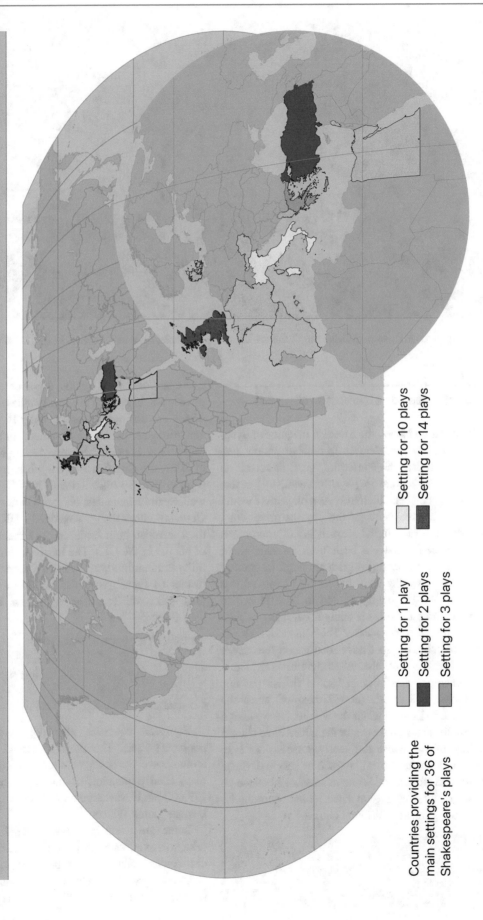

Countries providing the main settings for 36 of Shakespeare's plays

Setting for 1 play
Setting for 2 plays
Setting for 3 plays

Setting for 10 plays
Setting for 14 plays

Map 65

Fig. 47 Kronborg Castle in the town of Helsingør, Denmark—the setting of Hamlet's Elsinore. (Richard Mortel, CC BY 2.0)

66 Nobel Prize for Literature (Map 66)

Upon his death Alfred Nobel, the Swedish inventor of dynamite, left 94% of his considerable fortune to endow five annual awards for those who confer the "greatest benefit on mankind". The awards were in physics; chemistry; peace; physiology or medicine; and literature. As a physicist I would prefer to discuss the recipients of the physics prize, but I could not produce a meaningful map. It is not uncommon for a physics laureate to have been born in country A; educated in country B; done postgraduate work in country C; performed Nobel worthy work in country D; moved to a new, high-flying job in country E; and to have retired to country F by the time the prize is announced. It is not clear to me how to handle such a case. The situation with the literature prize is simpler, so I have focused on that. Even the literature prize is not always straightforward: Doris Lessing, for example, was born in Iran to British parents; her family moved to Rhodesia (now Zimbabwe) when she was 6; she moved to London when she was 30. Do we assign Lessing's prize to Iran, Zimbabwe, or the UK? I have chosen the UK; others might choose differently. Or consider Ivan Bunin, the 1933 laureate. He is classified as stateless. Since Bunin was born in Russia and wrote in Russian I believe he represents a win for Russia, and have tallied accordingly. Again, others might make a different choice.

> Home is not where you are born; home is where all your attempts to escape cease.—Naguib Mahfouz, 1988 Nobel Literature Prize Winner

Regardless of the details, a glance at the map shows the prize is not distributed evenly. Swathes of South America and Africa are not represented; Middle East and former Soviet countries are poorly represented; Indonesia, the world's fourth most populous country, has no literature laureate. As of 2021, the Swedish Academy has, however, awarded the prize to eight Swedish writers. Other European countries fare well. Spain: 5 laureates; Italy: 6; Germany: 9 (I give Hesse to Germany rather than Switzerland); UK: 11 (12 if we count the Tanzanian-born Abdulrazak Gurnah, who has spent most of his life in the UK). The most successful nation is France: Sully Prudhomme was awarded the first prize in 1901, and a further 14 French writers have been awarded it since. The most successful language is English, followed by French, then German.

> If there's a book that you want to read, but it hasn't been written yet, then you must write it.—Toni Morrison, 1993 Nobel Literature Prize Winner

European countries also dominate the award on a per-capita basis. The most successful place in per-capita terms, however, is tiny Saint Lucia (shown ringed on the map). The population of this West Indies island is only 177,000 but it was large enough to have produced the 1992 laureate Derek Walcott.

Some great authors missed out. Tolstoy, for heavens' sake, was overlooked. Joyce did not receive the prize. Neither did Auden, Borges, Chekhov, Frost, Kafka, Lawrence, Nabokov, Orwell.... One of my favourite authors, Graham Greene, was nominated 20 times but was always ignored

66 Nobel prize for literature

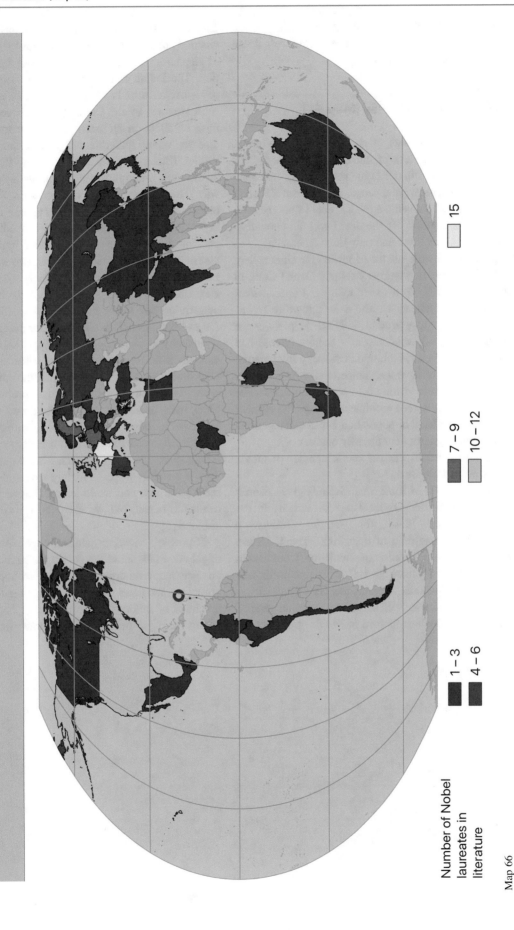

Number of Nobel
laureates in
literature

1 – 3
4 – 6
7 – 9
10 – 12
15

Map 66

(with rumours abounding that the Academician Artur Lundkvist blackballed Greene because of the author's affair with the Swedish actress Anita Björk).

67 Going to the Movies (Map 67)

During the family gatherings of my childhood, the adults would often lament the loss of cinemas. When *they* were young, they liked to recall, our home town had 13 cinemas. The growth in TV rentals (yes, people once rented TV sets) and then sales affected cinema attendance and thus the economic viability of cinemas. By the time I was old enough to 'go to the pictures' my town had only three cinemas.

I thought about this the last time I visited the cinema. My wife, who for some reason enjoys watching Daniel Craig in larger-than-life format, asked me to take her to see *Knives Out* (good film, recommended). The nearest town to us has just two cinemas to serve around 300,000 people. As we drove to one of them I wondered how many cinemas had closed here over the years; I thought how lucky my parents' generation had been to have the choice of so many venues. Until I realised: we were visiting a 14-screen multiplex. The other option was a six-screen multiplex. Perhaps the number of movie screens per head of population has not changed too much. And sound systems now are *way* better.

I wondered: how does the per capita number of movie screens vary around the world?

The tiny population of some small islands gives them a large ratio of screens to people. The four screens in the territory of Christmas Island, for example, produces a screen-to-million-people ratio of over 2000. The USA has just under 123 screens per million of population. Countries in western Europe typically have 50–100 screens per million of population—as do Australia, Canada, and New Zealand.

African countries are not well served with cinema screens. Nor is Saudi Arabia, which banned movie theatres between 1983 and 2018 (although the situation has changed and new cinemas are beginning to open). China has the largest number of cinema screens, but of course those serve a vast population. The Indian cinema industry produces more films than any other country, but has a relative paucity of screens.

A scar left by the Covid-19 pandemic might change all this. In 2020, in an attempt to slow transmission of the disease, many governments closed indoor entertainment venues. (Since people were reluctant to sit close to strangers in the dark, multiplexes would probably have closed anyway.) In response, film studios were forced to pull the blockbusters they had planned to release that year; much to my wife's regret, Daniel Craig's last outing as Bond was postponed several times. The situation improved in the summer of 2021, and people began a tentative return to cinemas. In September 2021 the new James Bond, *No Time To Die*, was finally released. Within days it became the best-performing box office hit since Covid-19 struck. But even James Bond has not saved the day for movie-industry bosses.

Industry insiders suggest *No Time To Die* will not recoup its costs from its run in movie theatres alone. It is viewers who stream the film, and watch it in the comfort of their own home, who will ensure the film is profitable. Many other films, even those that are well reviewed and in contention for awards, have been unable to attract crowds to the cinemas. It seems many adults remain wary of sitting in an indoor space, close to strangers, while a pandemic still burns. This is certainly the case for my wife and I: we travelled to see *Knives Out* and made an evening of it, just before the pandemic began, but we chose to watch *No Time To Die* at home.

If people's viewing habits change permanently then we might soon talk about the passing of multiplexes as a previous generation mourned the closing of 'the pictures'. Movie producers, who deal with projects that require investments of hundreds of millions of dollars, must hope the big screen retains its magic and its ability to draw people together.

67 Going to the movies

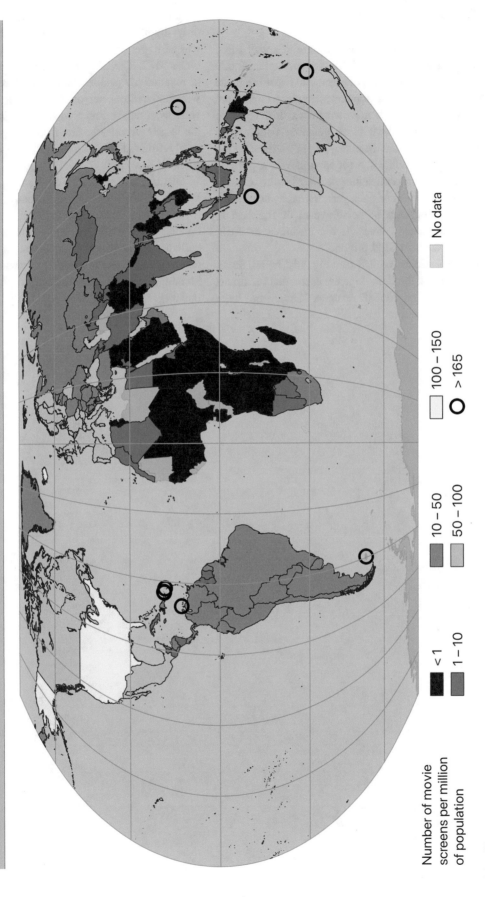

Number of movie
screens per million
of population

<1	10 – 50
1 – 10	50 – 100

100 – 150	No data
>165	

Map 67

68   Computer Games (Map 68)

Computer gaming is a part of global culture in a way I find difficult to understand. In July 2018, for example, 22,000 fans packed into a major New York venue to cheer the finals of the Overwatch League, a popular 'esports' competition. Although I would join such a crowd to watch a game of football or cricket, the thought of watching people play video games is alien to me. Similarly, I am immune to the attraction of popping bubbles on a smartphone, a pastime from which many of my colleagues derive pleasure. Clearly something is wrong with me: each year, humanity as a whole devotes tens of thousands of person-years of attention to these games. Computer gaming is big business: in 2019, global revenue was over $134 billion.

Six countries—China, the US (see Fig. 48), Japan, South Korea, Germany, and the UK—together accounted for almost three quarters of that global revenue. These are large countries, but even on a per-capita basis the story changes only slightly. Computer game revenue from Japan in 2019 was $140 per person; in South Korea the figure was slightly under $112. The US and UK came third and fourth in terms of revenue per person and Germany remained in the top ten with revenue per person at a similar level to Hong Kong, Norway, Canada, and Switzerland. Only China, which has the largest overall revenue from computer games, is significantly different when looked at in per-capita terms: it is on a par with Kuwait.

Gaming, and revenue from gaming, is not as widespread as, say, football. Gaming is absent in large parts of Africa; revenue in the Indian subcontinent is measured in cents not dollars; and gaming has little hold in South America. But perhaps the Covid-19 pandemic might change this. The pandemic forced the cancellation of many sporting events but home-based screen-based activity continued. Might 'esports' one day replace 'sports' in popularity?

68 Computer games

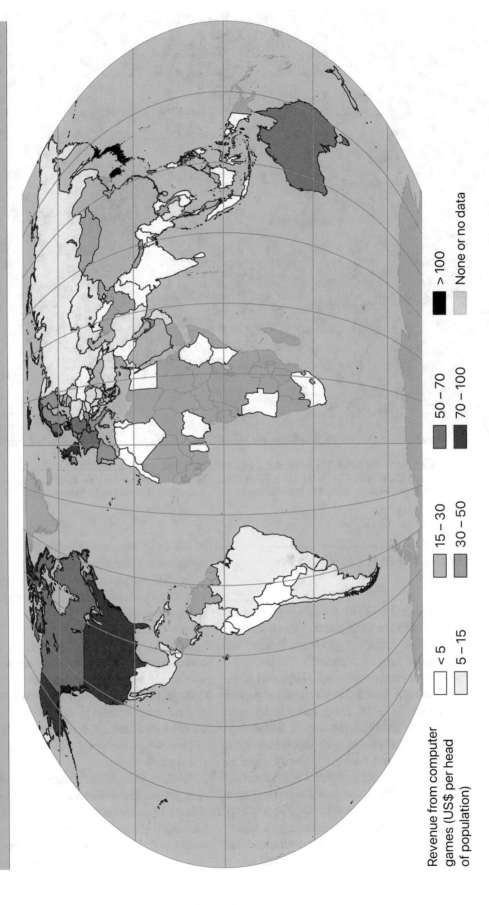

Revenue from computer
games (US$ per head
of population)

☐ < 5

☐ 5 – 15

■ 15 – 30

■ 30 – 50

■ 50 – 70

■ 70 – 100

■ > 100

☐ None or no data

Map 68

Fig. 48 In 2014, Seattle's KeyArena hosted The International (shown here), an annual esports world championship for the video game Dota 2. The total purse was $10 million. In 2021, the National Arena in Bucharest, Romania played host; the total purse had risen to $40 million. Might such events become more popular than football, baseball, or basketball? (Dota 2 The International, CC BY 2.0)

69   What the World Is Searching For (Map 69)

At the close of each year Google, the world's biggest search engine, publishes a list of the most popular search terms in countries for which it has records. It gives us a glimpse into what is on the collective mind of the world's citizens.

In 2018, football (or soccer) occupied a large part of the world's attention: the FIFA World Cup was the most popular search in 50 countries. In 2019, there was slightly more search variety but sport remained popular: in Bangladesh, India, and Pakistan, cricket was the most popular search; in Ireland and New Zealand it was the Rugby World Cup; in the UK it was the Rugby *and* Cricket World Cups; in several South American countries, the Copa América football competition came top. And non-sporting entertainment was popular in several countries: Americans searched about Disney Plus; Russians about Eurovision; several countries were interested in Thanos (a movie supervillain) or the final season of *Game of Thrones*. In some places, people searched for information about new social media platforms.

In 2018 and 2019, then, the commonest search theme in a majority of countries had to do with ways of passing the time: sport, entertainment, social media. And then the Covid-19 pandemic struck.

In 2020, entertainment was not so important. The most common search term in almost all nations was related in some way to the novel coronavirus sweeping across the planet. Only in the USA and one or two neighbouring countries was some other topic more important: Americans searched for news about the 2020 presidential election.

How would the world search in 2021, with the pandemic still raging?

Well, in many of the countries for which Google released data, people had reverted to a search for entertainment. The citizens of many nations searched for news of football: the Euro 2020 championship, postponed from the previous year because of the pandemic, was a focus for most European countries. Cricket was the main draw for India and

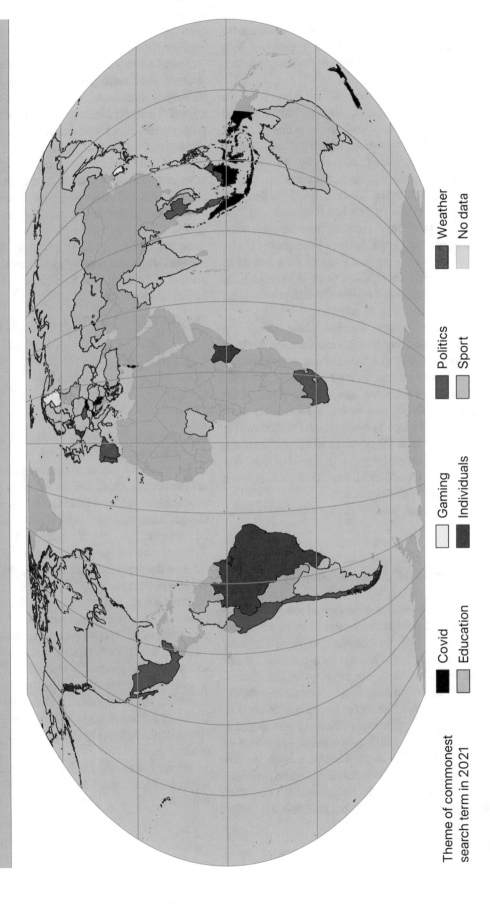

69   What the world is searching for

Theme of commonest
search term in 2021

Covid
Education

Gaming
Individuals

Politics
Sport

Weather
No data

Map 69

Pakistan. In other countries, basketball and the postponed Olympic Games (see Map 63) were the most searched-for topics. In Korea and Finland it was online games (see Map 68) rather than sports that drew people to the Google search engine.

Covid-related search terms were still the prime focus for Indonesia, Israel, New Zealand, Serbia, Slovakia, and Slovenia. In several other countries, internal policy decisions were important but these were often related to Covid and, in particular, the financial relief offered by the state to individuals affected by the pandemic. This was the case in Chile, Hong Kong, Malaysia, Mexico, South Africa, and Thailand. Only in Czechia (which had a census) and Peru (which had a close presidential election) were the main political stories unrelated to Covid. In three countries—Greece, Kazakhstan, and Turkey—the main search term related to online education. Again, though, this was Covid-related: the search terms were related to the move to online schooling, which the pandemic forced upon countries.

In four countries, people searched about people. In Belgium, the big news story was the manhunt for a soldier who had taken weapons from a barracks and wrote threatening letters aimed at virologists. In the Netherlands, the murder of an investigative journalist (see Map 18) captured the public's attention. In Brazil, fans wanted to know more about the airplane crash that killed a young, popular singer. And in Kenya, the death of John Magafuli, president of Tanzania, dominated search activity. Magafuli, a Covid-denier, was being treated in a Kenyan hospital, with rumours suggesting he had been infected by coronavirus.

In Spain and Portugal, weather was the main search topic: it was a *hot* year.

---

70   Active on Social Media (Map 70)

The year 2019 saw the 30th anniversary of the birth of the World Wide Web. The invention, developed by Tim Berners-Lee at CERN as a way for scientists to share research, changed the world. I was a relatively early adopter: I wrote my first html when the Web was just 4 years old. It took 16 years for the first billion users to go online and access the Web. In took just another 6 years for the number of users to reach 2 billion. Currently, the rate of increase is such that it takes just 2.7 years to add 1 billion new users. At the time of writing, in 2021, there are about 4.8 billion users so this current rate of increase cannot continue—soon everyone who wants to be connected will be connected. (The Web brought the internet to the masses, but note that the 'Web'

is just one aspect of the 'internet'. I suspect the two terms are synonymous for many, though, so here I do not distinguish between them.)

Since I first fired up a Mosaic browser the online world has changed, most notably in the widespread use of social media platforms—Facebook, YouTube, Instagram, TikTok, Twitter, etal. Three decades after the birth of the Web, 4.48 billion people were active social media users; collectively they spent hundreds of millions of hours on their preferred platforms. Much of that use happens on mobile devices, the possibility of which I presume Berners-Lee did not dream back in 1989. But this activity is not evenly distributed.

The starkest differences exist, as they do in so many ways, on the Korean peninsula. In South Korea internet use is essentially universal and 89.3% of the population is actively engaged with some form of social media. In North Korea, the internet is blocked (for its citizens, anyway; the Supreme Leader has access).

Global internet users: 4.80 billion.
Average daily time spent using the internet by each internet user: 6 hours 55 minutes.
Percentage of internet users accessing the internet via a mobile device: 92.1%.—Kemp (2021)

Unsurprisingly, in many parts of Africa social media use is not widespread. Only 0.2% of the Eritrean population is active on social media, a figure caused in part because of the president's restriction on all types of information flow (as discussed in Map 18). In Central African Republic, Chad, Congo, Malawi, Niger, and South Sudan fewer than 1 in 20 people use social media. Outside Africa, only Turkmenistan has such low usage (in that country, only 2.5% of the population use social media).

Of African countries, Libya stands out: following the fall of the Gadaffi regime, Libyans flocked to social media platforms to share and exchange information. Over two-thirds of the Libyan population are on Facebook.

In Brunei, Kuwait, Qatar, and UAE the use of mobile social media is essentially universal. Elsewhere the picture is one of countries gradually, but seemingly inevitably, becoming more connected. Yet another consequence of the Covid-19 pandemic might be an increase in social media use, with people being physically distant but wanting social connection.

We can even see the impact of social media on other worlds: right now, Perseverance Rover is tweeting from Mars.

70   Active on social media

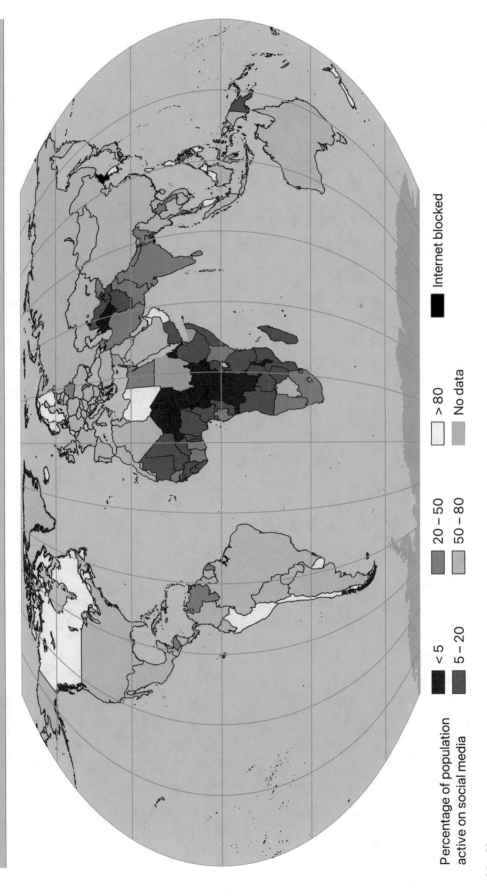

Percentage of population
active on social media

- < 5
- 5 – 20
- 20 – 50
- 50 – 80
- > 80
- Internet blocked
- No data

Map 70

The World of Economics

"Money makes the world go round", according to Liza Minelli and Joel Gray in the film *Cabaret*. Maybe. But sometimes the axles seize and the financial world stops turning. This happened during the financial crisis of 2008: queues formed at ATMs as savers rushed to withdraw their pounds, dollars, euros. It might happen again, as governments decide how to respond to the economic problems caused by Covid-19. The aftermath of the pandemic will no doubt play out differently in different countries. For example, countries that rely on international tourism suffered immediately when the world's pandemic response shut down air travel; industrialised countries that put workers on furlough are seeing their national debt soar—a fact that will presumably affect standards of living in future years. The maps in this chapter illustrate some of these aspects of money.

Of more importance than access to money, though, is access to the things that make life possible (such as food) or better (such as electricity) and the maps here also refer to these things.

We should note, though, that as our climate warms we might find none of the economic aspects discussed in this chapter really matter. There is no economy if all the people are dead.

71   Accessing Cash (Map 71)

I remember a time BA—before automated teller machines (ATMs, aka cashpoints or holes in the wall). To get money one used to have to write a cheque payable to cash; join a queue in one's bank (during bank opening times, of course); and, when your turn came, hand over the cheque to a teller. My parents have a friend who *still* uses this method of withdrawing cash. Personally, I can't imagine going back to the pain of writing cheques in order to swap it for banknotes.

I am sure few people ever give ATMs a moment's thought, unless they need wonga and are unable to find a cashpoint, but the effect of these devices on retail banking has been profound. Indeed, in 2009 Paul Volcker, Chairman of the Federal Reserve under Presidents Carter and Reagan, contrasted ATMs with the increasingly complicated financial instruments that had recently come to dominate the banking industry. ATMs had brought benefits to millions of people; those arcane financial instruments had brought the banking industry to the brink of disaster.

> The ATM has been the only useful innovation in banking.—Paul Volcker (2009)

It is impossible to point to a single invention and declare it to be the first ATM. Automatic dispensing technology for chocolate was available in 1950s USA; a machine for dispensing cash was trialled in 1966 in Japan; more robust trials took place the following year in the UK and Sweden, with the first public ATM being set up in June 1967 at a branch of Barclays Bank in Enfield, London. But for cashpoints to become mainstream required a combination of several innovations, both in engineering (keyboards and screens capable of withstanding weather and vandals; computer operating systems; banking cards; and so on) and in banking (PIN technology; a means of associating encrypted PINs with customer accounts; processes for restocking the dispensers; and so on). By the 1980s, ATMs were becoming a common sight in Europe and the USA.

The widespread adoption of the ATM led to a change in how high street banks interacted with their customers. The banks, no longing needing as many tellers, moved staff into selling products such as insurance, savings, and mortgages. When those financial products were unprofitable—well, the bank outlets themselves became less viable. Over time, the number of bank outlets has decreased. For my parents' friend, the act of cashing a cheque is becoming harder.

Macau has by far the highest number of ATMs per head of adult population (an impressive 325 per 100,000 adults).

© The Author(s), under exclusive license to Springer Nature Switzerland AG 2023
S. Webb, *Around the World in 80 Ways*, https://doi.org/10.1007/978-3-031-02440-5_9

71 Accessing cash

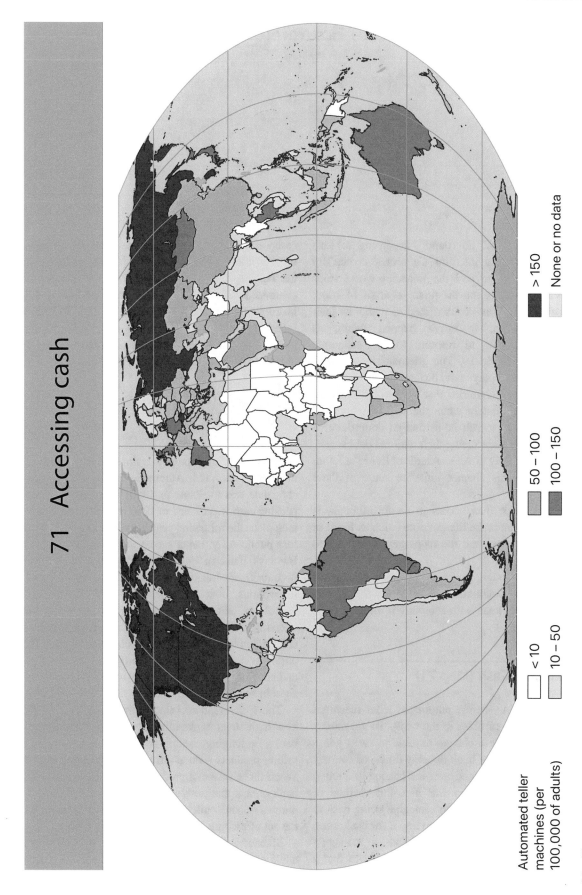

Automated teller
machines (per
100,000 of adults)

□ <10

■ 10 – 50

■ 50 – 100

■ 100 – 150

■ >150

□ None or no data

Map 71

Neither are you likely to struggle to find an ATM in South Korea (273); North America (Canada has 221, the USA has 153); Australia (169); Russia (161); and most European countries. You can even find cashpoints in the frozen wastes of Antarctica! McMurdo Station, the largest of the Antarctic research bases (see Map 1), has the continent's only ATMs. Workers at the station can get their cash from two ATMs. This might seem a paltry figure but, given that so few people are on the continent at any time, the number of cashpoints per head is surprisingly large. ATMs are far less common in African countries, in part because of the success of mobile banking in developing nations. Indeed, with the ubiquity of mobile technology and the growing interest in digital money I wonder whether the next generation will find ATMs—and cash itself—as old-fashioned as I find cheques.

72   Units of Currency (Map 72)

When clearing my loft, a chore that helped me pass some of the first Covid lockdown, I stumbled across my childhood collection of foreign coins. Cue an afternoon lost to Proustian flashback—the designs on those exotic groschen, paisa, and escudo had always led me to try and imagine the lives of the people who'd handled them. One notion I never thought to question, during my early numismatical phase, was this: the coins those people had spent as they went about their daily business carried less value than the coins I was using. I still recall the pre-decimalisation currency of my youth: £1 (then a banknote) contained 20 shillings and 1 shilling contained 12 pence; and I believed those British coins were 'worth more'. *Obviously* they were worth more: one pound sterling bought more than one unit of any other currency. That 'fact' made me proud.

A glance at the map shows this 'fact' still holds, more or less. On the day I checked, one pound sterling could buy you more than one unit of any other currency except that of Kuwait (£1 bought 0.40 dinar, the highest face value of any currency); Bahrain (0.50 dinar); Oman (0.51 rial); and Jordan (0.94 dinar). I realise now, of course, that this 'fact' has little meaning. The true Brit might as well feel proud that an inch is longer than a centimeter, an ounce weighs more than a gram, and a gallon container holds more liquid than a litre bottle. These are all just arbitrary units of measurement, with no deep significance.

Why, though, does the pound possess a high nominal value compared to most other units of currency? The answer dates back about 1300 years, when the Anglo Saxons decided to base their currency on silver and chose a pound of the metal as the unit of account. The pound of silver was rather impractical for everyday transactions but one could easily define smaller-valued coins. In 775, or thereabouts, the Anglo Saxons started to issue silver coins called 'sterlings', 240 of which could be minted from one pound of silver. It was only large payments that needed to be reckoned in 'pounds of sterlings', or 'pounds sterling' as it came to be known.

In 1816 the UK switched to the gold standard: £1 was valued at 7.3g of gold. For comparison, the US dollar at that time was valued at 1.6g of gold. So the nominal value of the pound sterling was still high. It retained a high value until World War II. The war effort, and subsequent economic decline, caused a floating pound to devalue. In recent decades it has stabilised—one pound will typically buy 1.2–1.3 US dollars.

Many currency units happen to be within a factor of 2 of the British pound. But so what? Strong economies can have 'small' currency units: £1 will buy you 1567 South Korean won, for example.

People need only worry about currency units when hyper-inflation strikes and confidence in money vanishes. In recent years the citizens of two countries in particular, Venezuela and Zimbabwe, have endured the ill effects of inflation. In Venezuela, in 2018, a cup of coffee cost 0.75 bolivars; in 2019, the same cup cost 2800 bolivars—and that was after a currency devaluation in 2018 that knocked five zeros off the currency. Many Venezuelans turned to cryptocurrencies because they had no faith in the bolivar. In Zimbabwe, between 2007 and 2008, a dose of hyperinflation set in and prices doubled every day. Banknotes with a face value of 100 trillion Zimbabwe dollars were in circulation. When the period of hyperinflation ended, the value of the Zimbabwean dollar had eroded so much that it was replaced with foreign currencies. The Zimbabwean dollar was reintroduced in 2019, and in the summer of 2021 the Zimbabwe government introduced its highest-denomination banknote: 50 dollars. That highest-value banknote was insufficient to buy a loaf of bread. The spectre of inflation is once again haunting that country.

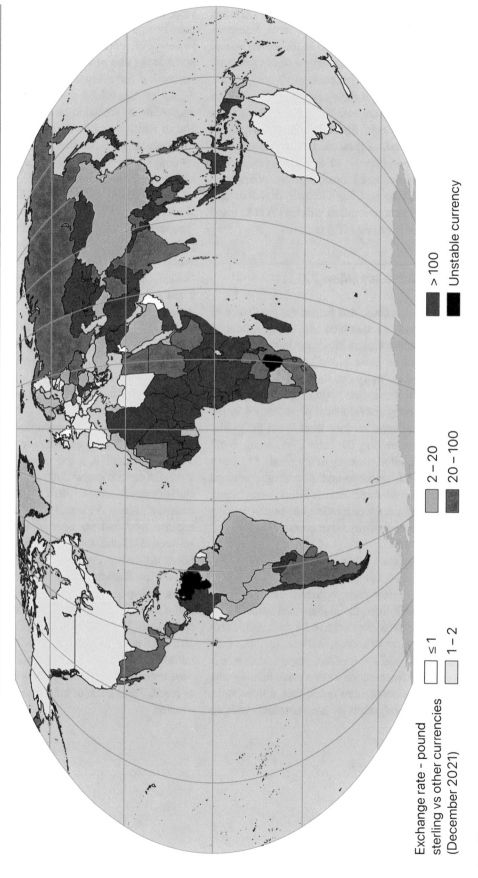

72 Units of currency

Exchange rate - pound
sterling vs other currencies
(December 2021)

☐ ≤ 1

░ 1 – 2

▨ 2 – 20

▨ 20 – 100

▨ > 100

■ Unstable currency

Map 72

73   Gold Reserves (Map 73)

The high-tech, polymer £10 note in my wallet carries the words "I promise to pay the bearer on demand the sum of ten pounds". A wrinkled, paper £20 note in my possession, which I have yet to exchange for the new polymer version, carries the same phrase, with "twenty" replacing "ten". Indeed, the wording "I promise to pay the bearer..." has appeared on English bank notes since the Bank of England began issuing them in 1694. But if I turn up to a bank, hand over my £10 note, and demand the promised ten pounds... what then happens? This question still worries me, one sign of many that I do not understand the concept of money.

> Although gold and silver are not by nature money, money is by nature gold and silver.—Karl Marx, in "Capital: The Process of Capitalist Production"

Back in the mists of time, in Anglo-Saxon England, a collection of 240 silver coins weighed one pound. That weight evolved into a currency, the pound sterling (see map 72). One could honour a written promise to pay the bearer a sum of money by swapping the promise for silver. Eventually, sterling and various other currencies moved to a gold standard. That move to gold made sense. From the mythical King Midas through to the present time people have valued this glittering, yellow metal—gold has been used as money for thousands of years. And, because gold is resistant to corrosion, it stands the best chance of any metal of maintaining its value. The metal, furthermore, is satisfyingly scarce: as Peter Bernstein pointed out in a book that traces humanity's obsession with the element, if you made a lump of all the gold ever mined then you could fit the lump on board a single oil tanker.

So could I turn up to a bank today and demand some gold in exchange for my banknote? Well, I could. But the bank would not give me gold.

The link with gold *used* to enable the Bank of England to maintain the value of its notes. But the Bank broke the link with gold back in 1931. Since then, there has been no particular asset—neither silver, gold, nor platinum—into which one has the right to convert bank notes. If I hand over my £10 note to a bank and demand my ten pounds then I will simply be given another £10 note (or the digital equivalent if the Bank of England develops Britcoin). The situation is the same in other countries. Money is more about faith than it is about assets, and public faith in a currency is maintained through a government's operation of the economy.

> Governments lie; bankers lie; even auditors sometimes lie: gold tells the truth.—William Rees-Mogg

In the event of an apocalypse, gold would be of only a little more practical use than paper notes. So I am not sure why some governments still choose to hoard gold. The USA has by far the world's largest reserves of gold, but on a per capita basis Switzerland is the outlier. For every citizen, the Swiss government holds 123 g of gold. On this basis the USA comes just tenth, with 25 g of gold per person. Seven of the top ten gold hoarders (on a per capita basis) are countries in western Europe. The other countries in the top ten, in addition to the USA, are Aruba and Lebanon.

73 Gold reserves

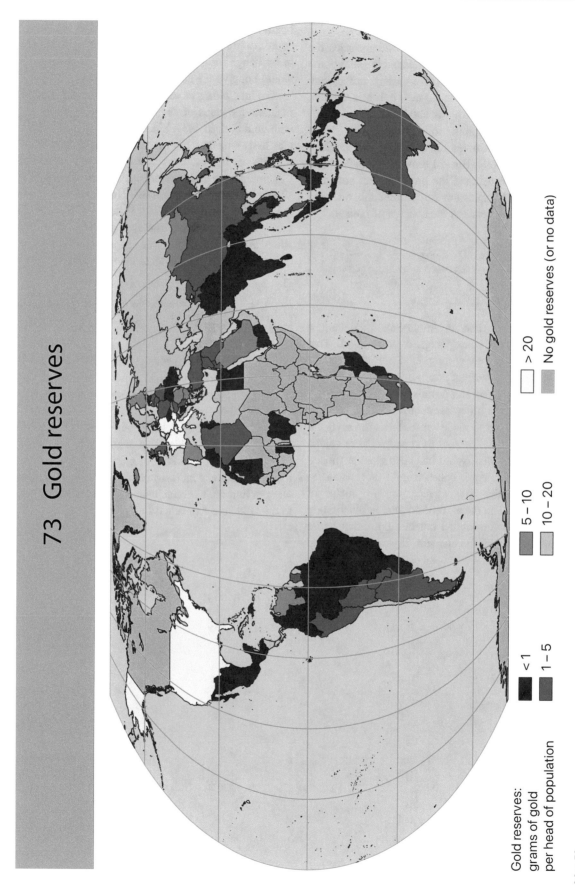

Gold reserves:
grams of gold
per head of population

▪ <1	▪ 5 – 10	□ > 20
▪ 1 – 5	▪ 10 – 20	▪ No gold reserves (or no data)

Map 73

74   The Indebtedness of Nations (Map 74)

News outlets, when analysing a country's economic situation, often focus on gross domestic product (GDP)—the total monetary value of all goods (cars, clothing, food...) produced in the country, plus the value of all services (banking, health care, education...) produced, plus the business investment required to generate that activity, all in a given time period (typically 1 year). As I argued in relation to Map 50, GDP is not a measure of wellbeing. But it *is* a widely quoted measure of a country's *economy*. The USA, China, and Japan have the biggest GDPs.

Those same news outlets, particularly since the financial crisis of 2008, often also refer to a country's debt. Debt is not inherently evil: an individual might take on debt to buy a house; a business might take on debt to invest in new machinery; a country might take on debt to finance a major infrastructure project. So a nation's total debt is not by itself too important. In 2017, for example, Greece's debt was $514 billion (it makes sense to use US dollars). Germany's debt was $2.7 trillion. But Germany's economy was more than 12 times bigger than Greece's. Germany bailed out Greece, not the other way round.

More important is the debt-to-GDP ratio, which investors use to gauge a country's ability to make future payments on its debt. News outlets usually give the ratio as a percentage, as I do here, but that is not quite right: GDP is a country's production in a given time period, so the true unit of debt-to-GDP is time. We can think of it as the number of years needed for a country to pay off its debt, if all its GDP went on repayment. In practice, since we all have to live, only a fraction of GDP can be devoted to debt repayment so the payoff time is longer.

The Covid-19 pandemic inflated the debt-to-GDP ratio of many countries: it shrank GDPs, while countries took on extra debt to respond to the health emergency. A World Bank study suggested that if the ratio stays above 77% for long periods, economic growth slows. So for many countries the future is perhaps less rosy than it seemed before the pandemic. Four countries have a debt at least two times bigger than GDP: Greece (206%), Sudan (259%), Japan (266%), and Venezuela (350%). Japan has long had a large debt-to-GDP ratio, but is unlikely to default. Venezuela, however, is vulnerable. See Fig. 49. Its economy has contracted by 75% since 2013, which is unprecedented in recent times for a peacetime economy. Its two main creditors are China and Russia, so this situation may develop a geopolitical aspect.

74 The indebtedness of nations

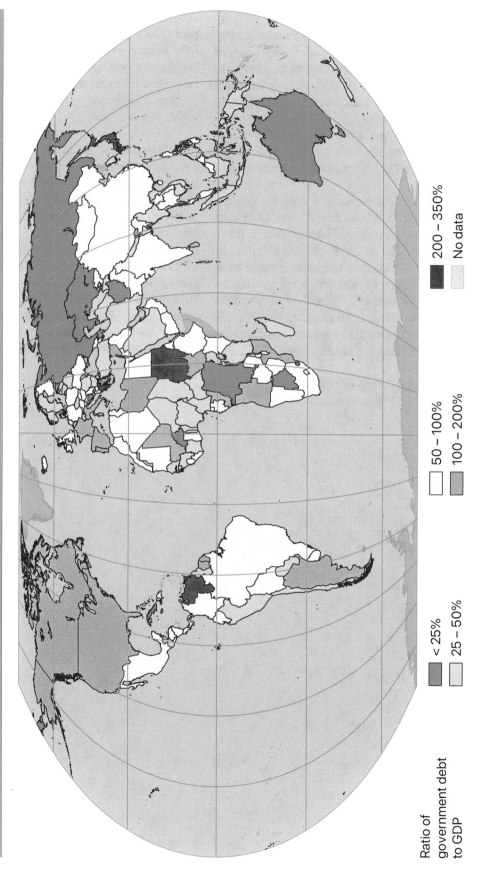

Ratio of
government debt
to GDP

■ < 25%
□ 25 – 50%
□ 50 – 100%
■ 100 – 200%
■ 200 – 350%
□ No data

Map 74

Fig. 49 Slums in Caracas. Over 5 million Venezuelans have fled the country, generating a refugee crisis for neighbours. (Public domain, CC0)

75   Standard of Living (Map 75)

As mentioned in Map 74, a country's gross domestic product (GDP), the yearly value of goods and services exchanged within its monetised economy, is only one measure of that country's economic health. It ignores wellbeing, for example, and the reality of people's lives. GDP is nevertheless widely and consistently measured, so GDP per capita is often used to compare standards of living. At least, it is used after making suitable adjustment for different prices in different countries, a concept that economists call purchasing power parity (PPP).

Suppose you buy a phone in London for £100 while the same phone in New York costs $120; the exchange rate would be $1.2 for every £1. (We assume here there are no other costs or barriers to trade.) Now, an exchange rate based on a single item is subject to a large margin of error so economists instead calculate a PPP exchange rate based on a large basket of goods. With the right exchange rate, customers in different countries have the same purchasing power—at least in theory. Comparisons of living standards in different countries are thus often made in terms of GDP per capita based on PPP. (The World Bank generates its figures using a hypothetical currency called the international dollar, which has the same purchasing power parity as the US$ has in the US at a particular point in time. The map shown here is based on analyses made using the international dollar.)

With that background out of the way we can ask the question: which places have the highest standard of living, defined in terms of GDP per capita?

Qatar is top, followed by Macau, Luxembourg, and Singapore. These all have a GDP per capita, based on PPP, in excess of $100K—but note that this tells us little about the distribution of wealth in those states. The Emir of Qatar is a billionaire; migrants working in construction projects for the 2022 FIFA World Cup are, according to Amnesty International, sometimes not paid for months. The densely populated gambling haven of Macau (see Maps 20 and 77) sees rich people spending vast sums in glitzy casinos but many more poor people living in cramped urban tenements. Inequality is an issue in both Luxembourg and Singapore.

The four places mentioned above all have small populations, and so the average can be pulled up by the presence of a few extremely wealthy individuals. The same is true of the tiny offshore financial haven of the Cayman Islands, a British Overseas Territory in the Caribbean.

Some larger countries with high living standards include the US; Australia; Saudi Arabia and other petrostates; and various Nordic and north European countries. In the next rank of wealth comes: Canada; New Zealand; Oman; Japan; South Korea; and many remaining European countries. No surprises there.

And of course there are no surprises when it comes to the poorest nations.

75　Standard of living

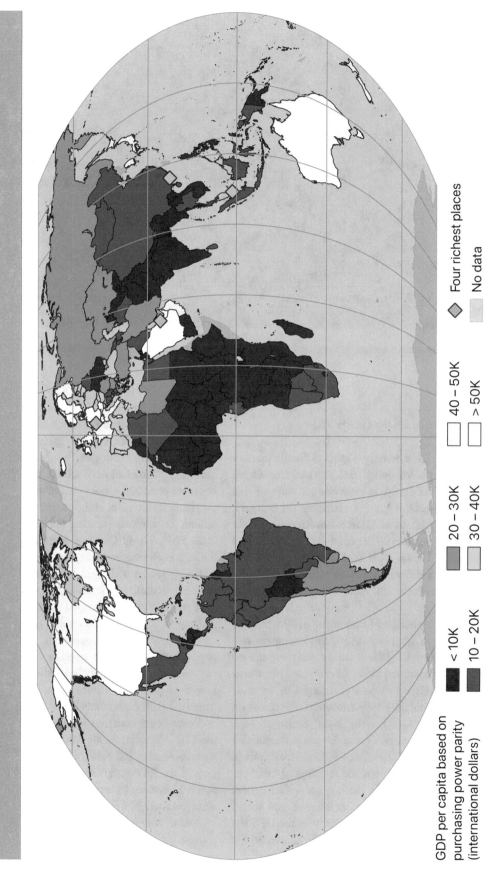

GDP per capita based on
purchasing power parity
(international dollars)

▮ < 10K	▮ 20 – 30K	☐ 40 – 50K
▮ 10 – 20K	☐ 30 – 40K	☐ > 50K

◈ Four richest places

☐ No data

Map 75

Three countries have a GDP per capita, based on PPP, below $10K: the Democratic Republic of the Congo; Central African Republic; and, poorest of all, Burundi. DR Congo is a big country, similar in area to western Europe. It enjoys large mineral deposits, excellent hydropower potential, and lots of arable land. And yet almost three quarters of the population live in poverty, as defined by the World Bank's international poverty rate. Directly north of DR Congo lies the Central African Republic. This country too possesses plenty of natural resources. A history of violent conflict (partly over access to those resources, partly down to disputes between different religious, ethnic, and political groups) has, however, locked its people into lives marked by low life expectancy (see Map 59) and low levels of literacy (see Map 46). To the east of DR Congo, and sharing a land border, is Burundi. More than 80% of the Burundi population live in poverty, as defined by the international poverty rate, while Map 79 shows how a low level of access to electrical energy blights the lives of many Burundi citizens.

76   Big Apple (Map 76)

In 1971, Steve Jobs (a college dropout) and Steve Wozniak (a self-taught electronics engineer) began a friendship. Four years later, they both attended a newly formed hobbyist group—the Homebrew Computer Club, in Menlo Park, California—at which Wozniak demonstrated an interesting device he had put together: a computer, attached to which was a keyboard and a monitor. The arrangement so impressed Jobs that he persuaded Wozniak to go into business with him and sell these devices. To fund the initial production, Jobs sold his VW campervan and Wozniak sold his electronic calculator (back then the calculator cost as much as the van). Together, they made a computer that eventually sold 200 units.

On 1 April 1976 Jobs and Wozniak founded Apple Computer Corporation. (The name 'Apple' seems to have come about because Jobs was then on a fruitarian diet, a diet he experimented with throughout his life. The legend of an apple falling on Newton's head, and inspiring a revolution in human thought, might also have played a role in the name: the company's original logo showed Newton sitting under a tree. It didn't harm, either, that in listings Apple appeared before Atari, a large computer firm at the time.) Jobs asked the businessman Ronald Wayne to be a co-founder of Apple

in return for a 10% stake. The idea was that Wayne could arbitrate if the two Steves fell out. Wayne agreed and founded Apple with Jobs and Wozniak, but backed out after just 12 days and sold his stake back for $500. Big mistake.

Wozniak's first Apple computer, which later became known as the Apple I, was the forefather of modern home desktop computers. Within months of the company's founding, Wozniak had designed an improved Apple II computer. Over the next few years the company grew. It soon faced competition, though, and by 1984, Apple was only a mid-size player in the personal computer market. The mighty IBM, a major supplier of computers to the corporate market, moved into personal computing and became a much bigger player than Apple. In 1985, with IBM seemingly invincible, Wozniak and Jobs left Apple. Wozniak left amicably; Jobs was forced out by Apple's CEO.

Over the next 12 years Jobs worked with his newly formed NeXT computing company, which was moderately successful. Apple, on the other hand, began a long decline. In early 1997 the company was within weeks of bankruptcy. In a last fling, Apple bought NeXT and brought Jobs back as an adviser. Within a few months of his return, Jobs was made CEO. As Apple's boss, an early master stroke from Jobs was to recognise the design genius of Jonathan Ive. Jobs, Ive, and Apple began to produce a string of easy-to-use consumer devices: the successful range of Macintosh desktop and laptop computers; iPods—portable digital audio players; iPads—tablet devices; and of course iPhones—smartphones that swept all before them. These devices changed the world.

The world so coveted these devices that, in August 2018, Apple became the first company to record a market capitalisation of $1 trillion. By late 2019 the company's market valuation exceeded the annual GDP of all but 14 countries (where GDP here refers to nominal GDP, rather than GDP adjusted for purchasing power parity as in Map 75). By August 2020, Apple's value exceeded $2 trillion. At the start of 2022, Apple became the first $3 trillion company. With the Covid-19 pandemic reducing the GDP of almost all nations, at the time of writing only four countries have a nominal GDP greater than Apple's market value: USA, China, Japan, and Germany.

One is tempted to say Apple's value cannot increase further. But the company is developing the Apple Car, an autonomous vehicle; and it is exploring augmented and virtual reality technologies. If Apple gets these right, the company could become even bigger.

76 Big Apple

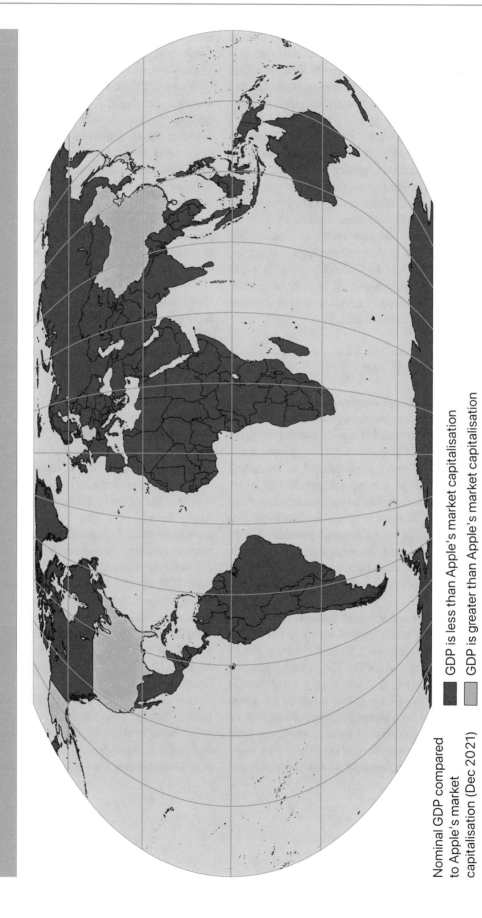

Nominal GDP compared
to Apple's market
capitalisation (Dec 2021)

■ GDP is less than Apple's market capitalisation

▨ GDP is greater than Apple's market capitalisation

Map 76

77   International Tourism (Map 77)

The flow of money, and of trades in goods and services, helps bind the world economy. International tourism is also a type of flow, a flow of people, which works in a similar way. Thus the World Bank, an international financial institution founded in 1944 to provide loans to the governments of low- and middle-income countries, is interested in tourism. According to the World Bank you are a tourist if you travel to a place outside of your usual environment, and stay there for no more than 1 year, for leisure, business, or any other activity for which you receive no payment from within the place you are visiting. Seems reasonable. But calculating the amount of international tourism is surprisingly tricky: the World Bank data, for example, differ from those of the World Travel and Tourism Council. Similarly, different approaches yield different figures for the total payments for goods and services received by a country from international tourists. The figures here happen to be based on World Bank data, but regardless of the methodologies used we can draw some general conclusions.

The last full year before the Covid-19 pandemic halted global tourism was 2019, and at the time of writing the World Bank has released tourism data for most countries for that year. (For the remaining countries, where data is collected at all, we have to be content with the data for 2018.) The figures show that, just before the pandemic brought international travel to a halt, the USA was the largest tourism economy in the world. International visitors from China were particularly important for the US tourism sector: although they made up only 4% of visitors they accounted for 11% of all spending. The vast US economy, however, has much bigger areas of activity than tourism. In a number of small countries and territories, receipts from international tourism constitute a significant fraction of total exports. Economic life in such places feeds on tourism.

Macau, a special Administrative Region of China, and the most densely populated region on Earth (see Map 20), is the place most reliant upon international tourism for its export economy. Gambling is the major draw here: casinos are illegal in mainland China and so the former Portuguese colony is a magnet for Chinese gamblers. Indeed, by revenue, Macau has become the gambling capital of the world.

In other places, beaches rather than betting bring in the visitors. The Maldives economy is almost as reliant on tourism as Macau's. It was not always that way: 50 years ago the Maldives depended on fisheries for its wealth. Cheaper air travel (see Map 23) led to a steady increase in visitors to the archipelago, and just before the pandemic struck the islands were hosting more than a million tourists each year.

A similar story holds for several Caribbean islands: Antigua and Barbuda, Grenada, Aruba, St. Lucia, St. Vincent and the Grenadines, and The Bahamas all rely heavily on international tourism for their export economy. Hotel resorts, scuba diving, ecotourism—these all bring in foreign currency. The other two nations with a large reliance on tourism are Palau, an archipelago in the western Pacific Ocean that consists of more than 300 coral and volcanic islands, and the equatorial São Tomé and Príncipe in the Gulf of Guinea.

But an economy that is heavily reliant upon one activity is vulnerable to shocks. In February 2020, Covid-19 forced Macau's casinos to close; during the summer of 2020, international flights no longer deposited pleasure seekers on the beaches of the Maldives. That year, the Maldives economy contracted by almost a third. Other island economies were similarly hit. Looking ahead, the impact of climate change on small island states will be severe: beach erosion (see Map 31), ecological damage, and the increased chance of storms will limit the appeal of tourism.

77 International tourism

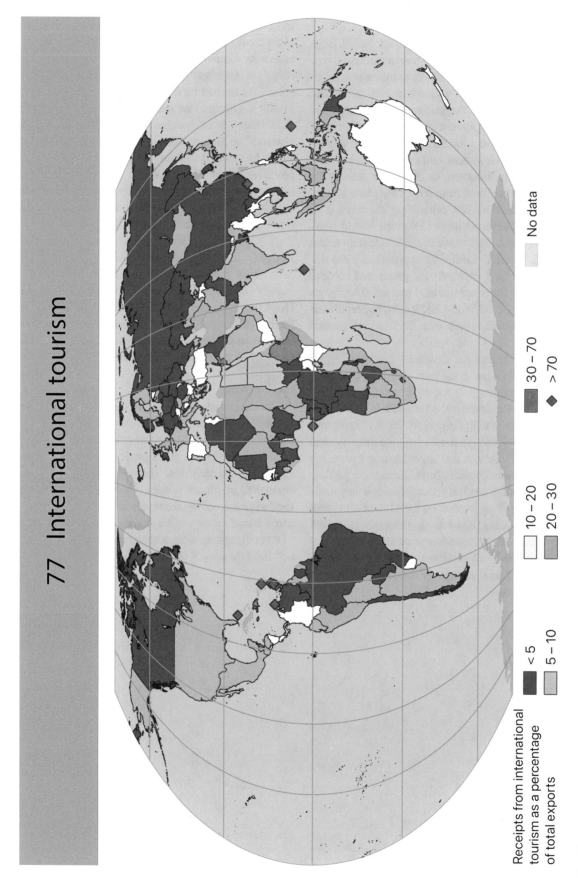

Receipts from international
tourism as a percentage
of total exports

■ < 5	□ 10 – 20	■ 30 – 70	▨ No data
▨ 5 – 10	▨ 20 – 30	◆ > 70	

Map 77

78 Military Expenditure (Map 78)

One of the key responsibilities of a national government is the security of the nation state itself. Every country understands the importance of defending itself against external threat. During wartime, military expenditure will absorb much of a country's economic output. In the twentieth century, for example, the fraction of the UK's economy devoted to military expenditure spiked twice—corresponding, of course, to the two World Wars. During peacetime, spending is generally smaller, but most countries still have deep pockets when it comes to defence. The Stockholm International Peace Research Institute (SIPRI) researches conflict, arms control, and disarmament. The SIPRI database shows that, immediately prior to the Covid-19 pandemic, most countries devoted at least 1% of their GDP to military spending. (The NATO definition of military spending includes all current and capital expenditure on any armed forces, along with spending on wages and pensions for personnel. Civil defence is excluded.)

Costa Rica and Panama were exceptions when it came to military spending. Neither country had a budget for its military in 2018. That year marked the 70th anniversary of Costa Rica's decision to get rid of its army; Panama abolished its army in 1994. Another interesting case was Japan. In 1944, Japan devoted almost 99% of its economy to the war effort; in 2018, it allocated only 0.92% of its GDP to spending on its military.

At the other end of the scale, five countries spent more than 5% of GDP on their forces. Four of them were in the Middle East: Saudi Arabia (8.8% of GDP), Oman (8.2%), United Arab Emirates (5.6%), and Kuwait (5.1%). The other country on the list was Algeria (5.3%).

In absolute terms the USA had by far the largest military spend. The US devoted 3.2% of its GDP to its military. In dollar terms, it spent more than the next nine nations combined.

One of the most concerning cases was hidden by 'No data'. Although its people live in grinding poverty, North Korea has one of the world's largest conventional military forces; see Fig. 50. It is testing nuclear weapons, too, along with the missiles needed to deliver them.

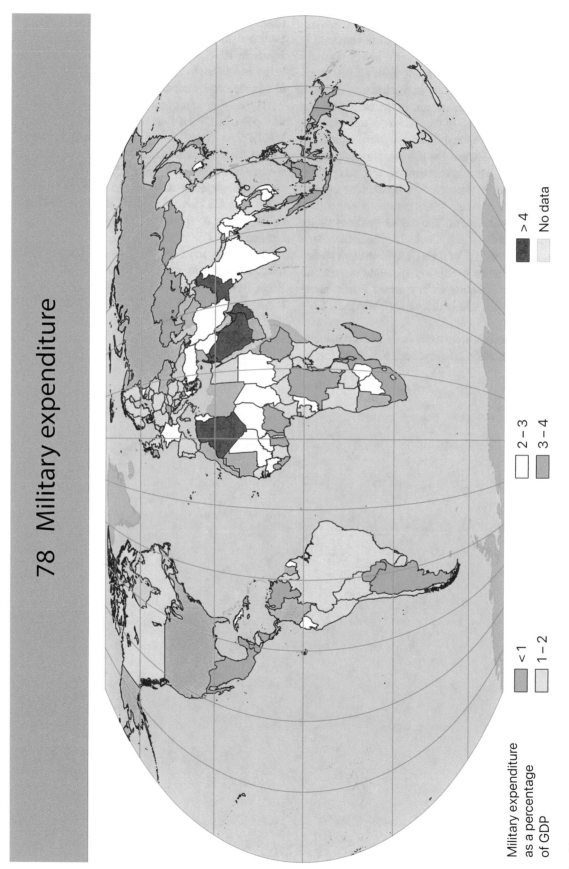

78 Military expenditure

Military expenditure
as a percentage
of GDP

< 1
1 – 2
2 – 3
3 – 4
> 4
No data

Map 78

Fig. 50 Thousands of North Korean soldiers marching in 2015 to celebrate the 70th anniversary of the Workers Party. North Korea is such a secretive state that no outsiders know what fraction of its GDP it devotes to the military, but the US State Department estimates the figure to be 25%. (Uwe Brodrecht, CC BY-SA 2.0)

79   Access to Electricity (Map 79)

I have a great capacity to take things for granted. My grandparents were born into a world in which basic aspects of life relied on muscle power. If you were rich, animals provided the muscles; if you were poor, you relied on your own muscles. Two generations later and the flick of a switch gives me access to almost miraculous levels of convenience. Only during a power cut do I notice my good fortune.

Imagine what would happen if—as happened in Venezuela in March 2019—most of your country was without electrical power for 5 days. Some people nowadays might imagine the worst-case scenario to be no internet. (Oh, the humanity.) But the disruption caused by a power outage can range from the costly (frozen food spoils) to the troublesome (pumps fail to get clean water into homes and dirty water away) to the potentially fatal (babies lose specialist neonatal care; elderly or disabled people cannot leave the upper levels of high-rise blocks; the dimming of traffic lights causes carnage in the streets, until traffic volume reduces because refuelling is not an option). Our privileged existence relies on us having access to electric power.

Under the International Energy Agency definition, a rural household lacks 'access to electricity' if it does not reach 250 kWh per year. For urban households the threshold is 500 kWh per year. In most countries, every citizen has access to electricity. Not every citizen has the same energy use: even in rich countries some people can afford only a basic level of access. But in a few countries—mainly in sub-Saharan Africa—many citizens are unable to access electricity, full stop. In Burundi, only 9.3% of the population have access to electricity; in Chad, 10.9%; in Malawi, 12.7%. Lack of access to electricity is a clear proxy for poverty.

For most of us lack of power comes as a shock. For Burundians, it is a way of life: children gather fuel instead of learning; adults burn biomass that results in air pollution; everyone suffers when the absence of coolers causes medicines and vaccines to perish, unused. See Fig. 51. Without access to electricity, Burundians will struggle to improve their standard of living.

We face a dilemma: how can we electrify the globe without exacerbating climate change?

79 Access to electricity

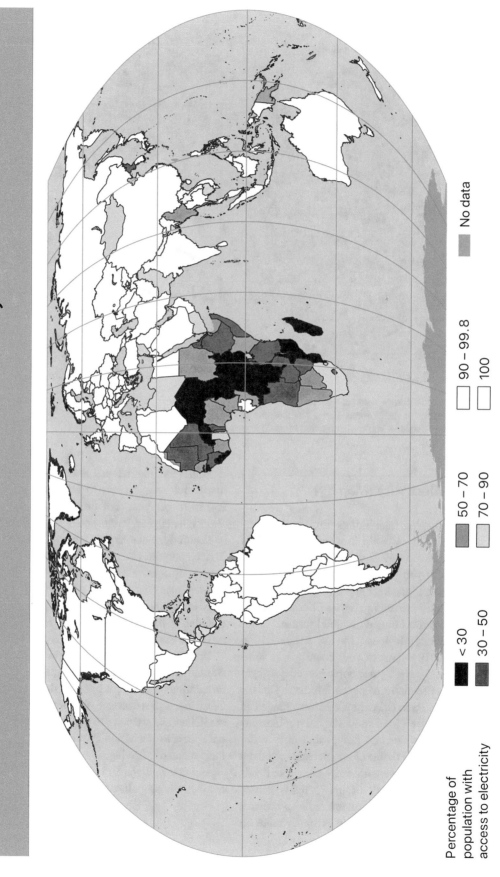

Percentage of
population with
access to electricity

■ < 30
■ 30 – 50
■ 50 – 70
□ 70 – 90
□ 90 – 99.8
□ 100
■ No data

Map 79

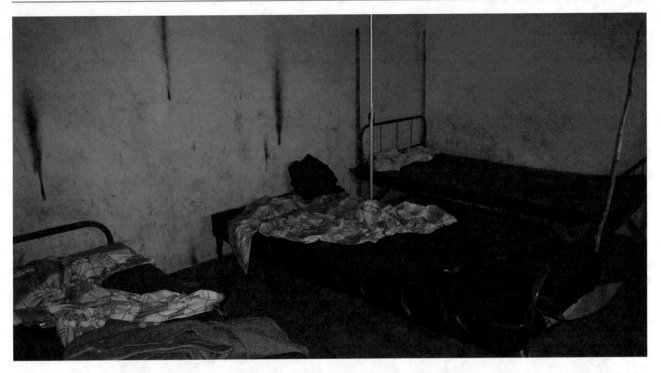

Fig. 51   A health center in Gitega, Burundi. In the absence of electricity, light is provided by kerosene lamps. Note how smoke from the lamps stains the walls. Pollution from kerosene lamps kills 10,200 Burundians every year. (© UNICEF Burundi/Carl Berndtsson)

80   Catching Fish (Map 80)

As a baby I was one of 3.2 billion mouths that needed feeding. The population is now 7.8 billion. In 2050, perhaps 10 billion people will be alive. How can Earth feed so many of us? And, given that animal agriculture produces more greenhouse gas emissions than planes, trains, and automobiles combined, how can we keep hunger at bay without torching the planet?

The UN's Food and Agriculture Organization, reviewing the state of the world's inland fishery resources, suggested that fishing in inland waters—lakes, rivers, ponds, canals, dams, and landlocked waters such as the Caspian Sea—can play a valuable role in feeding the world. In 2015, inland fisheries captured 11.47 million tonnes: 12.2% of the total global capture fishery production. That catch would be sufficient to meet the full dietary animal protein requirements of 158 million people. This is a vital source of high-quality nutrition.

Of course, this does not apply everywhere. The Arab region, and parts of north Africa, report no catch. In the UK, recreational fishing is more profitable than commercial eel net or trout and salmon fishing. But for people in some countries, inland fishing is a way of life.

Two countries have a particular reliance on inland fishing: Cambodia (which has an inland fishery catch of almost 35kg

per person per year) and Myanmar (just over 24kg). The rich aquatic biodiversity of the Mekong River system, which supplies the world's largest inland fishery, provides the Cambodian people with 80% of their animal protein. See Fig. 52.

After Cambodia and Myanmar, the next seven countries with the highest per capita catch of inland fish are all in Africa: Uganda, Chad, Congo, Malawi, Central African Republic, Mali, and Tanzania. Fishing contributes significantly to African food security.

Some of the world's poorest countries rely on inland fishery catch. They might have little other natural resource, but at least fisheries provide a source of animal protein that does not add to the problem of greenhouse gas emissions. These fisheries should be protected. And yet in many places they are under threat: a dam upstream can devastate fishing downstream; pollution can make rivers unusable; and climate change is drying up some lakes. We saw in Map 3 how the Aral Sea, once the fourth largest lake in the world and the site of a significant fishing industry, suffered an environmental catastrophe. As the lake shrank, so did the fishing community. With effort, and will, some of the worst depredations can be reversed. In the North Aral Sea, for example, there are indications that fishing is once again possible. But surely the wisest course of action is to protect these stretches of water from harm in the first place?

80 Catching fish

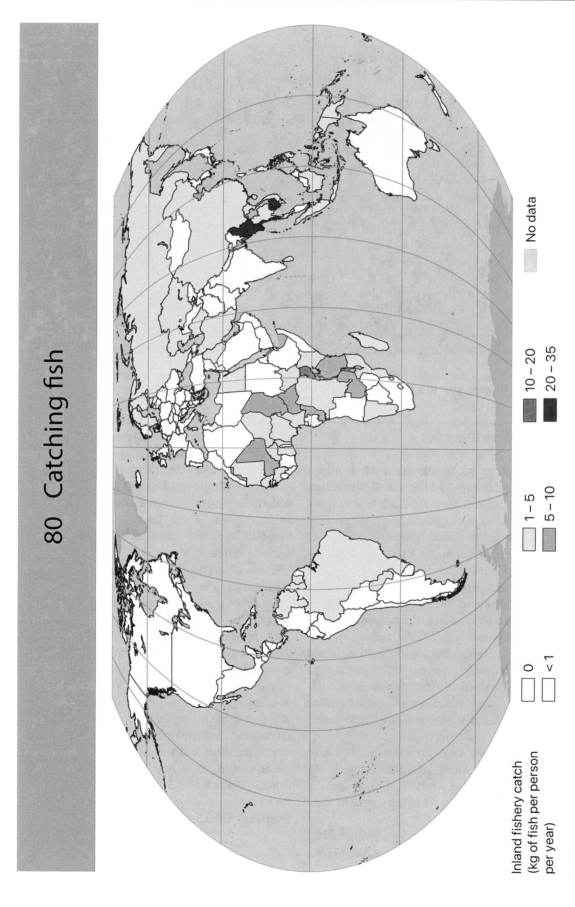

Inland fishery catch
(kg of fish per person
per year)

□ 0
□ < 1

■ 1 – 5
■ 5 – 10

■ 10 – 20
■ 20 – 35

■ No data

Map 80

Fig. 52 A woman fishing in Cambodia. People have fished here since ancient times, and they use more than 150 different types of angling equipment to catch hundreds of different types of fish. (Sylvain Raybaud, CC BY 2.0)

Appendix A: Afterword

On 19 July 2013, the Cassini spacecraft looked back through the rings of Saturn and took a photograph of Earth. (See Fig. 53.) If you look closely you can just about make out the Moon, too. From a distance of 1.44 billion kilometers Earth is just a dot with a blueish tinge. Every person then alive—indeed, as far as we know, every living creature in the universe—was on that dot.

An image such as this highlights an important message: we have only one planetary home. We are all in this together. If a medium-sized, fast-moving chunk of rock happened to collide with our blueish-tinged dot then it could spell the end for all of us. If the global climate emergency leads to a runaway effect then Earth will become uninhabitable for all people. If a new, virulent, airborne disease emerges in the wet markets of China then it will not pause at the country's borders before finding new hosts. We overestimate our importance if we believe the universe cares what happens on this small piece of cosmic property.

And yet what happens on Earth surely is important: as far as we know, human intelligence is the only example of advanced intelligence in the entire universe. So it is reasonable to look at how humans treat their home, how they choose to live, how they differ in their approach to planetary custodianship.

The maps highlight some stark differences between countries.

Perhaps the clearest contrast exists on the Korean peninsula, a strip of land that dangles from continental Asia and extends for roughly 1100 km—a trivial extent in the context of an image taken from Saturn, but large enough to contain almost two opposing worlds.

North Korea is the world's most secretive, tightly controlled, politically isolated country. Many of the maps in this book either have no data associated with North Korea or else data that are outdated or unreliable. The situation arises because the ruling Kim dynasty has for decades prioritised its own survival over the welfare of the North Korean people. No-one knows for sure what happens there since Kim Jong-un does not release accurate information, but economists estimate the annual output of North Korea is less than 1% of California's. The citizens of the country lack freedom; more than half of them lack access to electricity; several hundred thousand of them are believed to have starved to death in the 1990s. In many respects this is a failed country. But it maintains a fearsome military apparatus.

A 4-km-wide strip of land, a demilitarised zone running along the 38th parallel, separates North from South Korea. Where the North Korean economy is a basket case, the South Korean economy is one of the world's strongest. Where North Korea shuts itself off, South Korea is one of the world's most important export nations. South Korean companies are leaders in semiconductor technology and computer memory; they make some of the most recognisable consumer products on the planet; together they constitute one a trillion-dollar economy.

The Korean peninsula is small, about the size of the island of Great Britain, and yet it contains two countries with such different ways of life they may as well be in different solar systems.

Or consider the case of New Guinea, the second-largest island in the world (if we count Australia as a continent; Greenland is then the largest island). New Guinea is home to two nations. The western half of the island, called Western New Guinea, is part of Indonesia. The eastern half of the island is an independent country called Papua New Guinea. (The island's bisection arises because of the history of colonisation in the region. The western half formed part of the Dutch East Indies; the eastern half was colonised first by Germany, then by Britain, and finally by Australia before being granted independence.) Take a look at the maps and note how often the two halves of New Guinea have been assigned different colours.

Papua New Guinea is one of the world's least explored nations. It is home to more than three dozen uncontacted tribes, and the interior of the country is almost certainly a place of plant and animal species as yet unknown to science. Only about one person in seven lives in an urban setting; Port Moresby, the capital, has no road links to other towns—airplanes are a vital mode of transport. More than 800 languages are spoken in this country, making it the most linguistically diverse place on Earth. Many people

S. Webb, *Around the World in 80 Ways*, https://doi.org/10.1007/978-3-031-02440-5

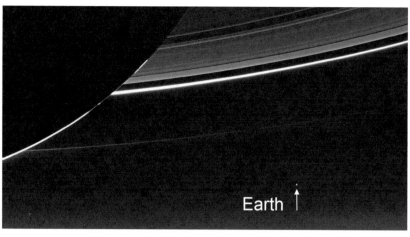

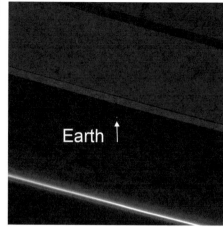

Fig. 53 Left: A view of Earth, just a tiny dot below the rings of Saturn, taken as part of the Cassini–Huygens mission, a joint project of NASA, ESA, and the Italian Space Agency. The world was told in advance that the Cassini spacecraft would take a planetary selfie. This was the first time this had happened, and many chose to celebrate. Did you wave to space? Right: Cassini's last view of Earth, taken in 2017, before the spacecraft burned up in Saturn's atmosphere. Our home planet is barely visible. (© NASA/JPL-Caltech/SSI)

here live in small farming groups, squeezing a sustainable existence from nature.

The contrast on these maps between the western and eastern halves of the island arises because, under the method used here, we are comparing a country where most of the population lives from semi-subsistence agriculture (Papua New Guinea) with a country that is the world's seventh largest in terms of GDP purchasing power parity (Indonesia). This does not mean the people in Western New Guinea are rich; far from it. The distribution of wealth in Indonesia is not uniform. But that east–west bisection of the island demonstrates, once again, how divergent human culture can be even in a limited area.

On a global scale we can see the disparities between nations when we compare the continent of Africa with China, the USA, Australia, or the countries of the EU.

When we do this, the first thing we realise is that Africa is vast. In terms of land area, Africa is about the same size as China *and* the USA *and* Australia *and* the countries of the EU. Africa covers a fifth of Earth's landmass. Given that the continent is so large, and that it is home to 1.2 billion people in 54 countries, variations inevitably exist within its boundaries. But these variations are less than those between Africa as a whole and the rest of the 'developed' world. Nowhere in Africa do the people have a life expectancy greater than 80 years, and in some African countries the main cause of death is easily preventable. Large numbers of people in some central African countries have little or no access to electricity, and on a GDP per capita basis they are poor. Apple's market capitalisation dwarfs the GDP of any country in Africa, but of course the majority of Africans cannot afford Apple products—and have poor internet access even if they could afford them. Africa is an unhappy place.

Diversity is surely a good thing. But inequality is dangerous. We have not yet learned to balance diversity and inequality. In a country as secretive as North Korea, which has an arsenal of perhaps 40 nuclear weapons and a stockpile of biological and chemical weapons, inequality could lead to a catastrophic war. The case of the island of New Guinea points to how inequality, if we are not careful, could lead to a widespread loss of linguistic, cultural and bio diversity. And Africa—well, per capita, the people of Africa emit small amounts of CO_2; the people of the developed world emit large amounts of CO_2. That inequality in CO_2 emission is doubly unfair: not only does it mean Africans are comparatively poorer, they are the ones who will bear much of the impact of the warming caused by CO_2 emission. The climate emergency has the potential to increase the severity of droughts in Africa, intensify heat waves, and cause crop failures—hundreds of millions of people might be forced to migrate. But where to? And what pressures might that migration bring to the world?

Hans Rosling in his book *Factfulness* painted an upbeat picture of the world. For some of us life is better now than it has been for people in the past. Far better to be alive today than at any other time in history. But there was nothing inevitable about the progress we have made and there is no guarantee we will continue to make progress, no guarantee the world will be a good place for our children or our grandchildren to live. Jump from a 30-storey building and you might enjoy the fall. It's never the fall that hurts; it's the stopping. The world's inequality might be what stops us.

Appendix B: Notes and Sources

B.1 The World Itself

Map 1 Earth's poles. NOAA (n.d.) has data on magnetic pole drift; Livermore et al. (2020) discuss why it drifts. NSIDC (n.d.) has sea-ice data. Asimov (1975) discusses polar exploration; Alexander (1998) focuses on Shackleton.

Map 2 Meteorites. For data, see The Meteoritical Society (n.d.). Bevan et al. (1985) discuss the Strathmore meteorite, Baxter and Atkins (1976) the Tunguska event. For meteorites in general, see Bevan and de Laeter (2002).

Map 3 Waterworld. See HydroLAKES version 1.0 (n.d.); for technical details, see Messenger et al. (2016). For more information about Earth's water, see Fagan (2011), Fishman (2012), Jha (2015), and Prud'homme (2011).

Map 4 Getting away from it all. Lukatela (n.d.) identifies Point Nemo; Stirone (2016) discusses how space agencies use it. Joshi and Cannon (1999) discuss Lovecraft's work and life. Butcher (1998) is a modern translation of Verne's novel.

Map 5 The longest straight line you can sail. For technical details of the branch-and-bound method used to calculate the longest straight line, see Chabukswar and Mukherjee (2018). For more about how Patrick Anderson initiated the initial internet discussion on this question, see Schultz (2018).

Map 6 The longest straight line you can walk without getting your feet wet. For Bruneau's suggestion, see IT/GIS Consulting (n.d.). The books mentioned are Asimov (1951–1953); Clarke (1973); Fermor (1977); Niven (1970); Pirsig (1974); Simon (1979); and Tolkein (1954–1955).

Map 7 Volcanic eruptions over the past 12,000 years. See Smithsonian Institution (n.d.). Brown (2018) has proximity data. For Snowball Earth, see Brocks et al. (2017); for Krakatoa, see Winchester (2003); for Tambora, see Klingaman and Klingaman (2014).

Map 8 Forests. See FAO (n.d., 2018) for data; consult Global Forest Watch (n.d.) for states not in the FAO database. Pearce (2018) discusses the uncertainty in deforestation figures.

Map 9 Chemical elements named after places on Earth. For information on the naming and discovery of the elements see, for example, Aldersley-Williams (2012), Emsley (2011), and Kean (2011).

Map 10 Earth's phosphorus. For details of phosphate reserves, see US Geological Survey (2020) and Kasprak (2016). For the phosphorus cycle, see Enger and Smith (2009). Mann (2018) and Vietmeyer (2012) discuss Borlaug's influence.

B.2 The World of Countries

Map 11 Countries named after people. For a thorough investigation of toponomy, the study of place names, see Everett-Heath (2005). The UN set up a working group to look at the use of exonyms and endonyms; see UNGEGN (2021).

Map 12 The gender gap. Data come from World Economic Forum (2021), which is the 15th edition of their Global Gender Gap Report.

Map 13 Female leaders. Worldometer (n.d.) provides world statistics. For biographies of some prominent female political leaders see, for example, Frank (2010); Ingham (2019); Qvortrup (2020); Seldon (2019).

Map 14 Flags containing red. See CIA (2020) for national flag data. Marshall (2017) describes the power national flags hold over some people. For the budding vexillologist, see Dorling Kindersley (2014), Grieg (2015), and Johnson (2015).

Map 15 Passport colours. Passport Index (n.d.) provides information on passport colour. See also the Henley Passport Index (2020). Both indexes provide information on the relative 'strength' of national passports.

Map 16 Driving—left or right? Kincaid (1986) discusses the history of the 'rule of the road'—the requirement for traffic to keep either to the right or to the left. See also Watson (1999).

Map 17 Capital punishment. Amnesty International (2021) global report provides information on the extent to which the death penalty is used by judiciaries across the world. See also Death Penalty Information Center; DPIC (n.d.).

Map 18 Press freedom. See RSF (n.d.) for access to the latest rankings on press freedom.

Map 19 Dominance of a capital city. See UN (2018) for data on world urbanisation. Bairoch (1988) presents the development of urbanisation, Brook (2014) the rise of new cities such as Dubai. Trantor, mentioned here, appears in Asimov (1951–1953).

Map 20 Population density. Many public sources provide data on territorial populations and surface areas (which here include inland bodies of water.) Where census data is unreliable, the UN Population Division publish estimates (UN, 2019).

B.3 The World Perturbed by People

Map 21 Carbon dioxide emissions. See Global Carbon Project (n.d.) for 2020 data. Berners-Lee (2019), Jaccard (2020), Klein (2015), Romm (2018), and Thunberg (2019) discuss the climate emergency and anthropogenic climate change.

Map 22 My short-haul carbon footprint. A report for the International Energy Agency provides details on the carbon dioxide emissions from the aviation industry; see Teter et al. (2019).

Map 23 Busiest airline routes. Overall statistics on the busiest routes are available from OAG (2019). See European Commission (n.d.) for information on options for reducing emissions from aviation.

Map 24 Renewable electricity. Data are from BP (2021) and Ember (n.d.). Maldonado (2017) discusses renewables in the EU's farthest regions. Mackay (2008) and Smil (2017) discuss the importance of energy generation for society.

Map 25 Nuclear power generation. PRIS (n.d.) is an IAEA database of nuclear power plants. See Mahaffey (2010) for a history of nuclear energy; Plokhy (2019) for Chernobyl; Lochbaum et al. (2014) for Fukushima.

Map 26 Trash. See Kaza et al. (2018) for details of the 'What a Waste' project.

Map 27 Plastic waste. See Kaza et al. (2018) for data. For more on the environmental problems associated with our use of plastic see for example Dorey (2018), McCallum (2019), and Siegle (2018).

Map 28 Pesticide use. See FAO (2019) for pesticide data; the fact that pests reduce annual global crop yields by up to 40% comes from FAO (2015). Carson (1962) warned of indiscriminate use of pesticides; Jameson (2019) revisits the topic.

Map 29 Mammal species under threat. See IUCN (2020) for data. For Indonesia, see Rainforest Action Network (n.d.); for Madagascar, see WWF (n.d.). Adams and

Carwardine (1990) and Carwardine (2009) detail species on the edge of extinction.

Map 30 Night sky brightness. Falchi et al. (2016) has data on night sky brightness. See Drake (2019) for the effect of lighting on human health, Sánchez-Bayo and Wyckhuys (2019) for the decline of insect species. See also IDA (n.d.).

B.4 The World of the Built Environment

Map 31 Homes at risk of rising sea levels. See Cohen and Small (1998), McGranahan et al. (2007). IPCC (n.d.) has data on rising sea level. Baxter (2005) discusses the North Sea flood. For information on the Delta Works see Deltawerken (n.d.).

Map 32 Lighthouses. Lighthouse locations are from OpenStreetMap (n.d.). See also Rowlett (2019) and ADLL (n.d.). For the history of lighthouses see Crompton and Rhein (2018), Grant (2018), Jones (2013), and Nancollas (2018).

Map 33 World heritage sites. UNESCO (n.d.) is an interactive list of World Heritage Sites. UNESCO (2018) is the eighth edition of a lavishly illustrated hard-copy guide to the sites.

Map 34 Roman amphitheatres. Bomgardner (2002) traces the origins and development of the Roman amphitheatre; Hopkins and Beard (2011) discuss this 'cathedral of death' in detail. Wilmott (2008) looks at Roman amphitheatres in Britain.

Map 35 The 100 tallest buildings. The Council on Tall Buildings and Urban Habitat (CTBUH) has developed internationally-accepted criteria for measuring the heights of tall buildings. See CTBUH (2021) for data on skyscraper location.

Map 36 Spaceports. See Roberts (2019) for data. Hall and Shayler (2001) and Brzezinski (2008) trace the Soviet space program. Spencer (2011) discusses Cape Canaveral. Iacomino (2019) tracks the role of private actors in space.

Map 37 The 500 most powerful supercomputers. Of all the maps in this book, this is likely to change most quickly. See TOP500 (n.d.) for the latest figures.

Map 38 Mains frequency. A number of websites monitor national grid frequencies in real time. For the UK grid, see e.g. Lawrence (2020). Cochrane (1985) is a history of the UK electricity grid; Thompson (2016) describes the US grid.

Map 39 Sharing a McDonald's. See McDonald's (n.d.) for data on store location; Kroc (2012) and Love (1995) detail the company's history. Langert (2019a, 2019b) explains how McDonald's responded to issues such as packaging, waste, and recycling.

Map 40 Soccer stadia. For data on grounds in England, see footballgroundmap.com (n.d.). Batchgeo (n.d.) gives the location of the 500 largest stadia in the world, which can be filtered by capacity and use.

B.5 The World of Ideas, Thoughts, and Beliefs

Map 41 The most complex machine ever built. See CERN (2020) and ATLAS Collaboration (2019). For more about particle physics see e.g. Butterworth (2016); Carroll (2013); and Gagnon (2018). For more about Higgs, see Nobel Prize (2013).

Map 42 Scientific productivity. The data here refer to the period 1 November 2018 to 31 October 2019. For further details see Nature Index (n.d.). See Nature (n.d.) for more background. For more on the Scimago project, see Scimago (n.d.).

Map 43 Intellectual property. See WIPO (n.d.) for details; the data here refer to 2018. Vaidhyanathan (2017) discusses the laws governing intellectual property and the values associated with how we disseminate ideas in a hyperconnected world.

Map 44 Views on the safety of vaccines. For more about the survey discussed in the text, see Wellcome Trust (2019). Rabesandratna (2019) explores some of its implications. For more on Dahl's attitude to vaccination, see Sturrock (2016).

Map 45 Good at maths? See PISA (2020) for all the reports on student performance. See OECD (1999) for background information regarding the PISA initiative. Niyozov and Hughes (2019) and Strauss (2019) provide a sceptical take on PISA.

Map 46 Literacy. CIA (2020) has data on literacy rates for most countries; World Bank (n.d.-a), basing its figures on the UNESCO Institute for Statistics, is another data source.

Map 47 Marking time. For more on calendars, see Asimov (1959) and Duncan (1999).

Map 48 UFO sightings. NUFORC (n.d.) points to a vast database of UFO sightings. For details of the sighting that initiated the 'flying saucer' craze, see Arnold (2014). Scoles (2020) looks at a social phenomenon that has persisted for several decades.

Map 49 How the world believes. Data are taken from the Pew Research Center databases; see Pewforum (2015) for estimates of the changing religious composition of countries up to the year 2050.

Map 50 The happiness of nations. Helliwell et al. (2020) is the latest World Happiness Report to which I have access. The report uses data from the Gallup World Poll; see Gallup (n.d.) and links therein.

B.6 The World of Wellbeing

Map 51 Homicide. Data come from the UN Office on Drugs and Crime's study on homicide; see UNODC (2019). Kleinfeld (2019) discusses security in the world's deadliest places.

Map 52 Suicide. Estimates come from the World Health Organization's Global Health Observatory data repository; see WHO (2021). See Helliwell et al. (2020) for details of the *World Happiness Report*.

Map 53 Obesity in adults. Data are from GHO (2020); health information is from CDC (2020). WHO (2010) looks at increasing levels of obesity in Pacific islanders. VNS (2018) reviews the twin issues of malnutrition and obesity in Vietnam.

Map 54 Alcohol consumption. See GHO (n.d.) for details on alcohol consumption per capita. Malhotra (2017) provides evidence for the suggestion that early humans drank alcohol. For details of the 'world's oldest brewery', found in present-day Israel, see BBC (2018).

Map 55 Road traffic deaths. WHO (2018) provides statistics on the toll taken by road traffic accidents. Glassbrook (2017) is an introduction to driverless cars.

Map 56 Incidence of malaria. Estimates on the incidence of malaria come from WHO (2019). Honigsbaum (2001), Webb (2009), and Rocco (2010) provide overviews of the history of malaria and the search for treatments and cures.

Map 57 Experience of pain yesterday. See Gallup (2019, 2021) for the two Global Emotions Reports mentioned in the text. Melzack and Wall (1996) and Wall (1999) are accounts of the biological causes and consequences of pain.

Map 58 How the world dies. WHO (2020) provides an overview on the commonest causes of death, and a link to mortality tables by country. Jauhar (2018) is an account of heart disease written by a cardiologist.

Map 59 The life expectancy gender gap. The UN Department of Economic and Social Affairs provides statistics on life expectancy at birth; see UN (2019). Shkolnikova (2016) discusses why men in Russia die so much earlier than women.

Map 60 Life expectancy at birth. The life expectancy statistics here are from the UN Department of Economic and Social Affairs; see UN (2019). For UK life-expectancy data, see ONS (2019). Austad (1997) is an introduction to why we age.

B.7 The World of Sport and Leisure

Map 61 Football World Cup. FIFA (n.d.) gives information on results in the World Cup. Connolly (2014) tells the story of West Auckland's triumph in the Sir Thomas Lipton Trophy; Clegg (1982) is a humorous dramatisation of the episode.

Map 62 Success at the Summer Olympics. Olympic Games (n.d.), the official Olympic website, links to a searchable database of results that can be filtered by Olympiad or by sport. Goldblatt (2018) is a history of the Games.

Map 63 Success at the Summer Paralympics. International Paralympic Committee (n.d.), the official Paralympic website, links to a searchable database of results that can be filtered by date, sport, and country.

Map 64 Chess playing strength. See FIDE (n.d.) for the ranking of chess-playing nations by average rating of the top 10 players. Figures here correct as of October 2021. For a recent history of the game, see Eales and Sloan (2019).

Map 65 Shakespeare's world. For further details on the settings of the Bard's plays, see introductions to the Arden Shakespeare; see also Folgerpedia (2016), one of many articles relating to the Folger Shakespeare Library.

Map 66 Nobel prize for literature. Nobel Prize (2021) gives details of all the literature laureates. Feldman (2000) and Worek (2011) discuss the history of the Nobel awards. See Allen (2007) for more on the literature prize in the period 1968–1980.

Map 67 Going to the movies. UIS (n.d.) and Statista (n.d.) have data on films and cinema. CinemaTreasures (2020), with details on over 53,000 movie theatres from around the world, claims to be the 'world's largest guide to movie theatres'.

Map 68 Computer games. For computer-game revenue data, see Newzoo (n.d.). See ATUS (n.d.) for data on time spent by Americans on gaming.

Map 69 What the world is searching for. GoogleTrends (2021) provides details of the top search terms in 2021, at both the country and thematic level. It provides a fascinating insight into the collective 'pulse' of the globe as a whole.

Map 70 Active on social media. Kemp (2021) has estimates of internet use across the world. See Hafner and Lyon (1998) for the origins of the internet; Berners-Lee (2000) and Gillies (2000) for the origins of the World Wide Web.

B.8 The World of Economics

Map 71 Accessing cash. World Bank (n.d.-b), based on the IMF's Financial Access Survey, has data on ATMs. Latest data for the USA come from Orem (2017). See Bátiz-Lazo (2015) for a history of ATMs.

Map 72 Units of currency. Latest exchange rates are available from any online currency converter. Ferguson (2019) is an account of the history of money; Eagleton and Williams (2007) give an overview of monetary systems and coinage.

Map 73 Gold reserves. BullionVault (2019) has details of the world's central bank gold reserves. Ferguson (2019) and Eagleton and Williams (2007) discuss the gold standard; Bernstein (2004) discusses the power of gold.

Map 74 The indebtedness of nations. Data on debt-to-GDP ratios are from TradingEconomics (n.d.). Note: CIA and IMF sources give slighly different numbers, but not enough to alter the map. See Map 75 and Deaton (2015) for more on GDP.

Map 75 Standard of living. See World Bank (n.d.-c) and CIA (2020). Coyle (2015) and Fioramonti (2013) discuss how GDP is calculated and the uses to which it is put. Raworth (2017) proposes an alternative metric to GDP.

Map 76 Big Apple. See World Bank (n.d.-e) for GDP data. Many books cover the story of the world's most valuable company. See for example Linzmayer (1993), Hertzfeld (2004), Kahney (2013), and Isaacson (2015).

Map 77 International tourism. For data on international tourism, see World Bank (n.d.-d). Becker (2016) examines how tourism, which encompasses airlines, hotels, and restaurants, became one of the world's major industries.

Map 78 Military expenditure. For data, see SIPRI (2020). Johnston (2017) and Sloan (2016) are two recent accounts of the history of NATO.

Map 79 Access to electricity. The SEforALL Global Tracking Framework, jointly led by the World Bank, International Energy Agency, and the Energy Sector Management Assistance Program, has data on electricity access. See SEforAll (n.d.).

Map 80 Catching fish. For data see Funge-Smith (2018), the UN report mentioned in the text. For further information on the Mekong, see OpenDevelopmentMekong (2018). For more on fishing in the North Aral Sea, see Chen (2018).

Appendix C: Technical Appendix

Each map in this book refers to a point in time. The maps will therefore age, some faster than others. You can, however, update them yourself. Not too long ago, the creation of a map of the world was an activity reserved for professional cartographers. Nowadays, if you have access to a moderately powerful computer and some basic computer skills then you can easily to it yourself. Furthermore, as I explained in the Introduction, you might disagree with how I have presented the data in these maps: again, you can create the maps for yourself and present the data in the way you believe makes most sense. I'd encourage you to do so, rather than rely on my interpretation.

Most of the maps in this book are choropleths, so in this appendix I present a method for creating a choropleth map of the world. The creation of a cartogram, as in Map 26, or an image-based plot, as in Maps 4–7, requires a slightly different approach. But if you can create a choropleth map of the world then you will have little difficulty in learning how to create one of those other forms of map.

To create a choropleth world map you need three things: a data set; a shapefile of the world; and a program for analysing the data and drawing the map.

C.1 Data Set

A data set is simply some number—such as literacy rate, life expectancy, or per capita CO_2 emission—associated with each country in the world. Many organisations make such information freely available. The UN Department of Economic and Social Affairs, for example, has a trove of openly accessible databases; visit population.un.or/wpp/ if you are interested in data on population and fertility, mortality and migration.

Once you have identified an interesting data set, download it in csv format (or else in a form that enables you to convert it to csv). A csv file, or comma-separated values file, is just a plain text file in which values are separated by a comma, a tab, or some similar delimiter. Each row in a csv file is a data record, with each record containing one or more fields. To create a choropleth you need a csv file with at least two fields:

one field contains the country identifier and a second field contains the number associated with that country (literacy rate, life expectancy, per capita CO_2 emission, or whatever it is you are investigating). If it helps, you can think of the csv file as being a table with two columns: in the first column is the name of the country, in the second the number associated with that country. Accessing a data set should be straightforward, but you need to bear in mind one subtlety: country names and codes.

Different data sets will have been generated by different people in different organisations following different standards, and there is little consistency about what certain countries should be called. Some refer to the Czech Republic, others to Czechia; some to Swaziland, others to Eswatini; some to the UK, others to the United Kingdom of Great Britain and Northern Ireland. Since you need to identify countries in order to generate a choropleth it makes little sense to use country *names*; instead, use the ISO two- or three-letter country *codes*. Your csv file should therefore contain two columns: the first is the country code, the second is the number of interest associated with that country. Note that, even when you use country codes, you should be prepared to encounter data inconsistency. For example, the code for South Sudan is SSD—but some sources use SDS; the code for Palestine is PSE—but some sources use PSX. The code value itself is unimportant; consistency is vital. So pick a value and make changes where necessary.

C.2 Shapefile

The second thing you need is a shapefile. A shapefile is a set of three or more files that contain geographical data. In particular, a shapefile contains information on geometric shapes (points, lines, and polygons), which a computer can knit together to generate the outline of a geographical object. A world shapefile contains shape data on the countries of the world, thus enabling a computer to depict the Earth. Because the shapes are represented by mathematical formulae, a computer can resize maps without losing resolution and it can easily apply a different map projections. The technicalities of the shapefile format are unimportant, but it might help to think of it as a large table of data. The table contains many columns, but only two are important for us here. One column contains the ISO two- or three-letter country code—which enables us to identify a country unambiguously and associate it with our number of interest (literacy rate, life expectancy, per capita CO_2 emission, or whatever it might be). Another column contains the shape data—which enables a computer program to draw the outlines of the world's countries.

So, where do you get a shapefile?

Natural Earth, an international, collaborative effort of cartographers, supported by the North American Cartographic Information Society, offers a public domain map dataset. To download a shapefile, visit naturalearthdata.com and follow the instructions.

C.3 Data Analysis and Visualisation

The third and final thing you need is some way of analysing your data set and presenting it on a choropleth. If you are a confident coder then you could quickly write a Python script, for example, to do generate a choropleth. For most people, though, it is easier to let geographic information system (GIS) software do all the work. Commercial GIS platforms are powerful, but can be expensive. Fortunately, there are a number of excellent open source alternatives—the most popular of which is probably QGIS.

The QGIS volunteer-driven project offers a free and open source geographic information system that lets you create, edit, and visualise geospatial information. It works on PCs, Macs, and Linux computers. You can download the version that is suitable for your computer by visiting qgis.org and following the 'For Users' tab.

* * *

With these three elements—a data set, a shapefile, and a program such as QGIS—you can create a choropleth in a few steps.

You can find many interactive lessons online, so I won't provide a full tutorial here. In any case, the details will differ depending on your computer and on the software version you downloaded. The following basic recipe should be enough to get you started, though.

Let's assume you have a data file called mydata.csv; a shapefile called earth.shp; and a working version of QGIS on your computer. Let's further assume mydata.csv contains two columns: one column holds the three-letter country codes (so ABW for Aruba all the way up to ZWE for Zimbabwe) and the other holds some number of interest (literacy rate, life expectancy, per capita CO_2 emission, or whatever) for each of the countries.

First, fire up QGIS and start a new project. Once the software is running, add the shapefile earth.shp as a layer in your project. Go to the Layer menu, choose Add Layer, and from there choose Add Vector Layer. (A shapefile is a type of 'vector file' because it stores its geographical information in a mathematical form rather than as an image.) Navigate to the directory where you installed earth.shp and choose the file. An image of the world should soon appear in the project window. You can zoom in, zoom out, change map projection—play with this a little so you get a feel for the program controls.

Second, add your data file as a separate layer. Go to the Layer menu, select Add Layer, and this time choose Add Delimited Text Layer. (Your csv file is a *text* file in which relevant values are separated, or *delimited*, by a comma.) If you are presented with an option at this point, select No geometry (attribute only table). (Some csv files might contain coordinates. You might, for example, want to plot the location of spaceports or volcanoes or capital cities. In this case you would be dealing with coordinates, and thus geometry. But for a choropleth you are mapping some *attribute* rather than a physical coordinate.)

So you now have a map of the world and a data set as two separate layers within QGIS. To generate a choropleth the two layers must somehow 'talk' to one another. The key that lets this happen is the three-letter country code: an identical column of data appears in both earth.shp and mydata.csv! If you join the attribute to the shapefile then QGIS can generate a choropleth. So: highlight the layer called earth.shp and select Layers then Layer Properties.... Choose the Joins option: you are going to 'join' your data file to the shapefile. Click the 'Add' or 'Plus' button; an Add Vector Join screen appears. On this screen ensure that the 'Join layer' is your data file; for the 'Join field' choose the column in which you have your three-letter country code; and for the 'Target field' choose the column in the shapefile that contains the three-letter country code—for the Natural Earth shapefile that column is called ADM0_A3.

You can now create your choropleth. Select the earth.shp layer and select 'Graduated' from the Layers Styling panel. The 'Value' you are plotting is the title of the data column in mydata.csv (literacy rate, life expectancy, CO_2 emission, or whatever). Choose the mode you want (equal count, equal interval, natural breaks; play with the options). Choose the number of different classes into which you want to put your data. Press Classify—and QGIS will generate a choropleth for you!

You can play with colours, modes, and class numbers. You can try different map projections. You can make a map that best illustrates your data. Just be aware, as discussed in the Introduction, of all the ways you might be fooling yourself.

References

Adams, D., & Carwardine, M. (1990). *Last chance to see*. Heinemann.

ADLL. (n.d.). *Admiralty digital list of lights*. www.admiralty.co.uk/publications/admiralty-digital-publications/admiralty-digital-list-of-lights

Aldersey-Williams, H. (2012). *Periodic tales: The curious lives of the elements*. Penguin.

Alexander, C. (1998). *Endurance: Shackleton's Legendary journey to Antarctica*. Bloomsbury.

Allen, S. (2007). *Nobel lectures in literature, 1968-1980*. World Scientific.

Amnesty International. (2021). *Death sentences and executions 2020*. Amnesty International.

Arnold, K. (2014). *The coming of the saucers*. CreateSpace.

Asimov, I. (1951–1953). *Foundation; foundation and empire; second foundation*. Gnome Press.

Asimov, I. (1959). *The clock we live on*. Abelard Schuman.

Asimov, I. (1975). *The ends of the earth: The polar regions of the world*. Weybright and Talley.

ATLAS Collaboration. (2019). *ATLAS: A 25-year insider story of the LHC experiment*. World Scientific.

ATUS. (n.d.). *American time use survey*. US Bureau of Labor Statistics. www.bls.gov/tus/

Austad, S. N. (1997). *Why we age: What science is discovering about the body's journey through life*. Wiley.

Bairoch, J. (1988). *Cities and economic development: From the dawn of history to the present* (C. Braider, Trans.). University of Chicago Press.

Batchgeo. (n.d.). *The 500 largest stadiums in the world*. https://blog.batchgeo.com/sports-stadium-capacities/

Bátiz-Lazo, B. (2015). A brief history of the ATM. *The Atlantic*. www.theatlantic.com/technology/archive/2015/03/a-brief-history-of-the-atm/388547/

Baxter, J., & Atkins, T. (1976). *The fire came by*. Doubleday.

Baxter, P. J. (2005). The east coast Big Flood, 31 January–1 February 1953: A summary of the human disaster. *Philosophical Transactions of The Royal Society A Mathematical Physical and Engineering Sciences, 363*. https://doi.org/10.1098/rsta.2005.1569

BBC. (2017, May 5). *Passports please*. BBC Radio 4.

BBC. (2018). *'World's oldest brewery' found in cave in Israel, say researchers*. www.bbc.co.uk/news/world-middle-east-45534133

Becker, E. (2016). *Overbooked: The exploding business of travel and tourism*. Simon & Schuster.

Berners-Lee, M. (2019). *There is no planet B: A handbook for the make or break years*. Cambridge University Press.

Berners-Lee, T. (2000). *Weaving the web: The original design and ultimate destiny of the world wide web*. HarperBusiness.

Bernstein, P. L. (2004). *The power of gold*. Wiley.

Bevan, A., & de Laeter, J. (2002). *Meteorites: A journey through space and time*. Smithsonian.

Bevan, A. W. R., Hutchison, R., Easton, A. J., Durant, G. P., & Farrow, C. M. (1985). High possil and strathmore – A study of two L6 chondrites. *Meteoritics, 20*, 491–501.

Bomgardner, D. L. (2002). *The story of the Roman amphitheatre*. Routledge.

BP. (2021). *Statistical review of world energy* (70th ed.). BP.

Brocks, J. J., Jarrett, A. J. M., Sirantoine, E., Hallmann, C., Hoshino, Y., & Liyanage, T. (2017). The rise of algae in Cryogenian oceans and the emergence of animals. *Nature, 548*, 578–581.

Brook, D. (2014). *A history of future cities*. Norton.

Brown, S. (2018). *Hawaii volcano: How many people do volcanoes kill?* www.bbc.co.uk/news/world-us-canada-44212666

Brzezinski, M. (2008). *Red moon rising: Sputnik and the rivalries that ignited the space age*. Bloomsbury.

BullionVault. (2019). *World's central bank gold reserves and interactive map*. www.bullionvault.com/gold-news/infographics/worlds-central-bank-gold-reserves-interactive-map

Butcher, W. (Trans.). (1998). *Jules Verne – Twenty thousand leagues under the seas* (Oxford World's Classics). OUP.

Butterworth, J. (2016). *Most wanted particle: The inside story of the hunt for the Higgs, the heart of the future of physics*. Experiment.

Carroll, S. (2013). *The particle at the end of the universe: How the hunt for the Higgs Boson leads us to the edge of a new world*. Dutton.

Carson, R. (1962). *Silent spring*. Houghton Mifflin.

Carwardine, M. (2009). *Last chance to see: In the footsteps of Douglas Adams*. Collins.

CDC. (2020). *Adult obesity: Causes & consequences*. www.cdc.gov/obesity/adult/causes.html

CERN. (2020). *ATLAS*. https://atlas.cern/

Chabukswar, R., & Mukherjee, K. (2018). *Longest straight line paths on water or land on the Earth*. https://arxiv.org/pdf/1804.07389.pdf

Chen, D.-H. (2018). Once written off for dead, the Aral Sea is now full of life. *National Geographic* www.nationalgeographic.com/news/2018/03/north-aral-sea-restoration-fish-kazakhstan/

CinemaTreasures. (2020). *Your guide to movie theatres*. http://cinematreasures.org/

CIA. (2020). *The world factbook*. US Govt Printing Office (also available from www.cia.gov).

Clarke, A. C. (1973). *Rendezvous with Rama*. Gollancz.

Clegg, T. (Director). (1982). *The world cup: A captain's tale*. Tyne Tees Television.

Cochrane, R. (1985). *Power to the people: History of the electricity grid*. LBS.

Cohen, J. E., & Small, C. (1998). Hypsographic demography: The distribution of human population by altitude. *The Proceedings of the National Academy of Sciences, 95*, 14009–14014.

Connolly, M. (2014). *The Miners' triumph: The first English world cup win in football history*. Oakleaf Publishing.

Coyle, D. (2015). *GDP: A brief but affectionate history*. Princeton University Press.

Crompton, S. W., & Rhein, M. J. (2018). *The ultimate book of lighthouses: An illustrated companion to the history, design, and lore*. Chartwell.

CTBUH. (2021). *Advancing sustainable vertical urbanism*. www.ctbuh.org

Deaton, A. (2015). *The great escape: Health, wealth, and the origins of inequality*. Princeton University Press.

Deltawerken. (n.d.). *The delta works*. www.deltawerken.com/Deltaworks/23.html

Dorey, M. (2018). *No. More. Plastic.: What you can do to make a difference*. Ebury.

Dorling Kindersley. (2014). *Complete flags of the world: The ultimate pocket guide*. Dorling Kindersley.

DPIC. (n.d.). *Death penalty information center – About us*. https://deathpenaltyinfo.org/about/about-us

Drake, N. (2019). Nights are getting brighter, and Earth is paying the price. *National Geographic*. www.nationalgeographic.co.uk/space/2019/04/nights-are-getting-brighter-and-earth-paying-price

Duncan, D. E. (1999). *Calendar: Humanity's epic struggle to determine a true and accurate year*. Harper Perennial.

Eagleton, C., & Williams, J. (2007). *Money: A history*. British Museum Press.

Eales, R., & Sloan, S. (2019). *Chess: The history of a game*. Ishi Press.

Elhacham, E., Ben-Uri, L., Grozovski, J., Bar-On, Y. M., & Milo, R. (2020). Global human-made mass exceeds all living biomass. *Nature*. https://doi.org/10.1038/s41586-020-3010-5

Ember. (n.d.). *Data*. www.ember-climate.org/data

Emsley, J. (2011). *Nature's building blocks: An A–Z guide to the elements*. OUP.

Enger, E. D., & Smith, B. (2009). *Environmental science: A study of interrelationships*. McGraw Hill.

European Commission. (n.d.). *Reducing emissions from aviation*. https://ec.europa.eu/clima/policies/transport/aviation_en

Everett-Heath, J. (2005). *Concise dictionary of world place-names*. OUP.

Fagan, B. M. (2011). *Elixir: A history of water and humankind*. Bloomsbury.

Falchi, F., Cinzano, P., Duriscoe, D., Kyba, C. C. M., Elvidge, C. D., Baugh, K., Portnov, B. A., Rybnikova, N. A., & Furgon, R. (2016). The new world atlas of artificial night sky brightness. *Science Advances, 2*. https://doi.org/10.1126/sciadv.1600377

Feldman, B. (2000). *The nobel prize: A history of genius, controversy and prestige*. Arcade.

Ferguson, N. (2019). *The ascent of money: A financial history of the world*. Penguin.

Fermor, P. L. (1977). *A time of gifts*. John Murray.

FIDE. (n.d.). *Federations ranking*. https://ratings.fide.com/topfed.phtml

FIFA. (n.d.). *Home page*. www.fifa.com/

Fioramonti, L. (2013). *Gross domestic problem: The politics behind the world's most powerful number*. Zed.

Fishman, C. (2012). *The big thirst: The secret life and turbulent future of water*. Simon & Schuster.

FAO. (n.d.). *Forest land*. www.fao.org/faostat/en/#data/GF

FAO. (2015). *Keeping plant pests and diseases at bay: Experts focus on global measures*. www.fao.org/news/story/en/item/280489/icode/

FAO. (2018). *Global forest products: Facts and figures*. UN FAO.

FAO. (2019). *Pesticides indicators*. www.fao.org/faostat/en/#data/EP

Folgerpedia. (2016). *List of settings for Shakespeare's plays*. https://folgerpedia.folger.edu/List_of_settings_for_Shakespeare%27s_plays

Footballgroundmap.com. (n.d.). *Football grounds*. www.footballgroundmap.com/grounds

Frank, K. (2010). *Indira: The life of Indira Nehru Gandhi*. HarperCollins.

Funge-Smith, S. (2018). *Review of the state of the world fishery resources: Inland fisheries*. FAO Fishery and Aquaculture Department.

Gagnon, P. (2018). *Who cares about particle physics?: Making sense of the Higgs Boson, the Large Hadron Collider and CERN*. OUP.

Gallup. (n.d.). *How does the Gallup World Poll work?* www.gallup.com/178667/gallup-world-poll-work.aspx

Gallup. (2019). *Global emotions report*. www.gallup.com/analytics/248906/gallup-global-emotions-report-2019.aspx

Gallup. (2021). *Global emotions report*. https://www.gallup.com/analytics/349280/gallup-global-emotions-report.aspx

GHO. (n.d.). *Global Information System on Alcohol and Health (GISAH)*. www.who.int/gho/alcohol/en/

GHO. (2020). *Global Health Observatory (GHO) data: Overweight and obesity*. www.who.int/gho/ncd/risk_factors/overweight_text/en/

Gillies, J. (2000). *How the web was born: The story of the world wide web*. OUP.

Glassbrook, A. (2017). *The law of driverless cars: An introduction*. Law Brief.

Global Carbon Project. (n.d.). *Home page*. www.globalcarbonproject.org/

Global Forest Watch. (n.d.). *Forest monitoring designed for action*. www.globalforestwatch.org/

Goldblatt, D. (2018). *The games: A global history of the olympics*. Pan.

GoogleTrends. (2021). *Year in search 2021*. https://trends.google.com/trends/yis/2021

Grant, R. G. (2018). *Sentinels of the sea: A miscellany of lighthouses past*. Thames & Hudson.

Grieg, C. (2015). *The directory of flags: A guide to flags from around the world*. Ivy Press.

Hafner, K., & Lyon, M. (1998). *Where wizards stay up late: The origins of the internet*. Simon & Schuster.

Hall, R., & Shayler, D. J. (2001). *The rocket men: Vostok and Voskhod, the first Soviet manned spaceflights*. Springer-Praxis.

Helliwell, J. F., Layard, F., Sachs, J., & De Neve, J.-E. (Eds.). (2020). *World happiness report 2020*. Sustainable Development Solutions Network.

Henley Passport Index. (2020). *Passport index*. www.henleypassportindex.com/passport

Hertzfeld, A. (2004). *Revolution in The Valley: The insanely great story of how the Mac was made*. O'Reilly Media.

Honigsbaum, M. (2001). *The fever trail – The hunt for the cure for malaria*. Macmillan.

Hopkins, K., & Beard, M. (2011). *The colosseum*. Profile.

HydroLAKES Version 1.0. (n.d.). *Overview*. www.hydrosheds.org/page/hydrolakes

Iacomino, C. (2019). *Commercial space exploration: Potential contributions of private actors to space exploration programmes*. Springer.

IDA. (n.d.). *Light pollution*. www.darksky.org/light-pollution/

IEA. (2018). *Renewables*. www.iea.org/fuels-and-technologies/renewables

Ingham, B. (2019). *The slow downfall of Margaret Thatcher: The diaries of Bernard Ingham*. Biteback.

International Paralympic Committee. (n.d.). *Paralympic and para sport results*. www.paralympic.org/results

Investopedia. (2020). *At $1.3 trillion, Apple is bigger than these things*. www.investopedia.com/news/apple-now-bigger-than-these-5-things/

IPCC. (n.d.). *AR1: Scientific assessment of climate change*. www.ipcc.ch/report/ar1/wg1/sea-level-rise/

Isaacson, W. (2015). *Steve Jobs: The exclusive biography*. Simon & Schuster.

IT/GIS Consulting. (n.d.). *Longest distance on land*. https://sites.google.com/site/guybruneau/fun-stuff/longest-distance-on-land

IUCN. (2020). *The IUCN Red List of threatened species*. www.iucnredlist.org/

Jaccard, M. (2020). *The citizen's guide to climate success: Overcoming myths that hinder progress*. Cambridge University Press.

Jauhar, S. (2018). *Heart: A history*. Oneworld.

Jameson, C. M. (2019). *Silent spring revisited*. Bloomsbury Wildlife.

Jha, A. (2015). *The water book.* Headline.

Johnston, S. A. (2017). *How NATO adapts: Strategy and organization in the Atlantic Alliance since 1950.* Johns Hopkins University Press.

Johnson, B. (2015). *The complete guide to flags of the world.* IMM.

Jones, R. (2013). *Lighthouse encyclopedia: The definitive reference.* Globe Pequot.

Joshi, S. T., & Cannon, P. (1999). *More annotated H.P. Lovecraft.* Bantam.

Kahney, L. (2013). *Jony Ive: The genius behind Apple's greatest products.* Penguin.

Kasprak, A. (2016). The desert rock that feeds the world. *The Atlantic.* www.theatlantic.com/science/archive/2016/11/the-desert-rock-that-feeds-the-world/508853/

Kaza, S., Yao, L. C., Bhada-Tata, P., & Van Woerden, F. (2018). *What a Waste 2.0: A global snapshot of solid waste management to 2050.* World Bank.

Kean, S. (2011). *The disappearing spoon ... and other extraordinary true tales from the periodic table.* Black Swan.

Kemp, S. (2021). *Digital 2021: global overview report.* https://datareportal.com

Kincaid, P. (1986). *Rule of the road: An international guide to history and practice.* Greenwood.

Klein, N. (2015). *This changes everything: Capitalism vs. the climate.* Penguin.

Kleinfeld, R. (2019). A path to security for the world's deadliest countries. *TED Summit 2019.* www.ted.com/talks/rachel_kleinfeld_a_path_to_security_for_the_world_s_deadliest_countries#t-135932

Klingaman, E. K., & Klingaman, N. P. (2014). *The year without summer: 1816 and the volcano that darkened the world and changed history.* St Martin's Griffin.

Kroc, R. (2012). *Grinding it out: The making of McDonald's.* St Martin's Paperbacks.

Langert, B. (2019a). *The battle to do good: Inside McDonald's sustainability journey.* Bingley.

Langert, B. (2019b). The business case for working with your toughest critics. *TED Summit 2019.* www.ted.com/talks/bob_langert_the_business_case_for_working_with_your_toughest_critics

Lawrence, J. S. T. (2020). *GB mains frequency.* http://mainsfrequency.uk/fm-home

Linzmayer, O. W. (1993). *Apple Confidential 2.0.* No Starch Press.

Livermore, P. W., Finlay, C. C., & Bayliff, M. (2020). Recent north magnetic pole acceleration towards Siberia caused by flux lobe elongation. *Nature GeoScience.* https://doi.org/10.1038/s41561-020-0570-9

Lochbaum, E., Lyman, E., & Stranahan, S. Q. (2014). *Fukushima: The story of a nuclear disaster.* New Press.

Love, J. F. (1995). *McDonalds: Behind the arches.* Bantam.

Lukatela, H. (n.d.). *Point Nemo (or, one thousand and four hundred miles from anywhere).* www.lukatela.com/pointnemo/index.html

Mackay, D. J. C. (2008). *Sustainable energy – Without the hot air.* UIT Cambridge.

Mahaffey, J. (2010). *Atomic awakening: A new look at the history and future of nuclear power.* Pegasus.

Maldonado, E. (2017). *Final report: Energy in the EU outermost regions.* https://ec.europa.eu/regional_policy/sources/policy/themes/outermost-regions/pdf/energy_report_en.pdf

Malhotra, R. (2017). *Our ancestors were drinking alcohol before they were human.* www.bbc.co.uk/earth/story/20170222-our-ancestors-were-drinking-alcohol-before-they-were-human

Mann, C. C. (2018). *The wizard and the prophet: Two remarkable scientists and their dueling visions to shape tomorrow's world.* Knopf.

Marshall, T. (2017). *Worth dying for: The power and politics of flags.* Elliott & Thompson.

McCallum, W. (2019). *How to give up plastic: Simple steps to living consciously on our blue planet.* Penguin Life.

McDonald's. (n.d.). *Your right to know.* www.mcdonalds.com/gb/en-gb/help/faq.html

McGranahan, G., Balk, D., & Anderson, B. (2007). The rising tide: Assessing the risks of climate change and human settlements in low elevation coastal zones. *Environment and Urbanization, 19,* 17–37.

Melzack, R., & Wall, P. (1996). *The challenge of pain.* Penguin.

Messager, M. L., Lehner, B., Grill, G., Nedeva, I., & Schmitt, O. (2016). Estimating the volume and age of water stored in global lakes using a geo-statistical approach. *Nature Communications, 13603.*

Nancollas, T. (2018). *Seashaken houses: A lighthouse history from Eddystone to Fastnet.* Particular.

Nature. (n.d.). *Nature.* www.nature.com

Nature Index. (n.d.). *Nature Index.* www.natureindex.com

Newzoo. (n.d.). *Key numbers.* https://newzoo.com/key-numbers

Niven, L. (1970). *Ringworld.* Ballantine.

Niyozov, S., & Hughes, W. (2019). Problems with PISA: Why Canadians should be skeptical of the global test. *The Conversation.* https://theconversation.com/problems-with-pisa-why-canadians-should-be-skeptical-of-the-global-test-118096

NOAA. (n.d.). *Wandering of the geomagnetic poles.* www.ngdc.noaa.gov/geomag/GeomagneticPoles.shtml

Nobel Prize. (2013). *The Nobel prize in physics 2013.* www.nobelprize.org/prizes/physics/2013/summary

Nobel Prize. (2021). *All Nobel prizes in literature.* www.nobelprize.org/prizes/lists/all-nobel-prizes-in-literature

NSIDC. (n.d.). *About NSIDC and the cryosphere.* https://nsidc.org/about

NUFORC. (n.d.). *The national UFO reporting center.* www.nuforc.org

OAG. (2019). *Busiest routes 2019.* www.oag.com/reports/busiest-routes-2019

OECD. (1999). *Measuring student knowledge and skills: A new framework for assessment.* OECD Publications. www.oecd.org/education/school/programmeforinternationalstudentassessmentpisa/33693997.pdf

OECD. (2006). *PISA released items – Mathematics.* www.oecd.org/pisa/38709418.pdf

Olympic Games. (n.d.). *Homepage.* www.olympics.com/en

ONS. (2019). *National life tables, UK: 2016 to 2018.* www.ons.gov.uk/peoplepopulationandcommunity/birthsdeathsandmarriages/lifeexpectancies/bulletins/nationallifetablesunitedkingdom/2016to2018

OpenDevelopmentMekong. (2018). *Fishing, fisheries and aquaculture.* https://opendevelopmentmekong.net/topics/fishing-fisheries-and-aquaculture

OpenStreetMap. (n.d.). *About OpenStreetMap.* www.openstreetmap.org/about

Orem, T. (2017). US ATM count hits half-million mark. *Credit Union Times.* www.cutimes.com/2017/08/11/u-s-atm-count-hits-half-million-mark-2

Palin, M. (1989). *Around the world in 80 days with Michael Palin.* [TV series] BBC.

Passport Index. (n.d.). *Explore the world of passports by country.* www.passportindex.org

Pearce, F. (2018). Conflicting data: How fast is the world losing its forests? *Yale Environment360.* https://e360.yale.edu/features/conflicting-data-how-fast-is-the-worlds-losing-its-forests

Pewforum. (2015). *Religious composition by country, 2010-2050.* www.pewforum.org/2015/04/02/religious-projection-table/2010/percent/all

Pirsig, R. M. (1974). *Zen and the art of motorcycle maintenance: An inquiry into values.* William Morrow.

PISA. (2020). *PISA.* www.oecd-ilibrary.org/education/pisa_19963777

Plokhy, S. (2019). *Chernobyl: History of a tragedy.* Penguin.

Premier League. (n.d.). *Home page.* www.premierleague.com/

PRIS. (n.d.). *Nuclear share of electricity generation in 2018.* https://pris.iaea.org/PRIS/WorldStatistics/NuclearShareofElectricityGeneration.aspx

Prud'homme, A. (2011). *The Ripple effect: The fate of fresh water in the twenty-first century*. Scribner.

Qvortrup, M. (2020). *Angela Merkel: Europe's most influential leader*. Duckworth.

Rabesandratna, T. (2019). France most skeptical about science and vaccines, global survey finds. *Science*. www.sciencemag.org/news/2019/06/france-most-skeptical-about-science-and-vaccines-global-survey-finds

Rainforest Action Network. (n.d.). *Indonesia's rainforests: Biodiversity and endangered species*. www.ran.org/indonesia_s_rainforests_biodiversity_and_endangered_species/

Raworth, K. (2017). *Doughnut economics*. Cornerstone.

Roberts, T. G. (2019). *Spaceports of the world*. Center for Strategic and International Studies.

Rocco, F. (2010). *The miraculous fever-tree: Malaria, medicine and the cure that changed the world*. HarperCollins.

Romm, J. (2018). *Climate change: What everyone needs to know*. Oxford University Press.

Rosling, H., Rosling, O., & Rosling Rönnlund, A. (2018). *Factfulness: Ten reasons we're wrong about the world – And why things are better than you think*. Hodder & Stoughton.

Rowlett, R. (2019). *The lighthouse directory*. www.ibiblio.org/lighthouse

RSF. (n.d.). *2021 World Press Freedom Index*. https://rsf.org/en/ranking_table

Sánchez-Bayo, F., & Wyckhuys, K. A. G. (2019). Worldwide decline of the entomofauna: A review of its drivers. *Biological Conservation, 232*, 8–27.

Šavrič, B., Patterson, T., & Jenny, B. (2018). The equal earth map projection. *International Journal of Geographical Information Science, 33*(3), 454–465.

Schultz, D. (2018). This ocean path will take you on the longest straight-line journey on Earth. *Science*. https://doi.org/10.1126/science.aau0354

Scimago. (n.d.). *Country rankings*. www.scimagojr.com/countryrank.php

Scoles, S. (2020). *They are already here: UFO culture and why we see saucers*. Pegasus.

SEforAll. (n.d.). *Data and evidence*. www.seforall.org/data-and-evidence

Seldon, A. (2019). *May at 10*. Biteback.

Shkolnikova, S. (2016). *PassBlue – Independent coverage of the UN*. www.passblue.com/2016/10/15/in-russia-rampant-alcoholism-among-men-leaves-women-with-rotten-choices/

Siegle, L. (2018). *Turning the tide on plastic: How humanity (and you) can make our globe clean again*. Trapeze.

Simon, T. (1979). *Jupiter's travels*. Hamilton.

SIPRI. (2020). *Stockholm International Peace Research Institute databases*. www.sipri.org/databases

Sloan, S. R. (2016). *Defense of the West: NATO, the European Union and the transatlantic bargain*. Manchester University Press.

Smil, V. (2017). *Energy and civilization: A history*. MIT Press.

Smithsonian Institution (n.d.). *Global volcanism program*. https://volcano.si.edu

Spencer, D. D. (2011). *Cape canaveral: America's spaceport*. Schiffer.

Statista. (n.d.). *Radio, TV & film*. www.statista.com/markets/417/topic/476/radio-tv-film/

Stirone, S. (2016, June 13). This is where the International Space Station will go to die. *Popular Science*.

Strauss, V. (2019). Expert: How PISA created an illusion of education quality and marketed it to the world. *Washington Post*. www.washingtonpost.com/education/2019/12/03/expert-how-pisa-created-an-illusion-education-quality-marketed-it-world

Sturrock, D. (2016). *Storyteller: The life of Roald Dahl*. William Collins.

Teter, J., Le Feuvre, P., Gorner, M., & Scheffer, S. (2019). *Tracking transport: Aviation*. www.iea.org/reports/tracking-transport-2019/aviation

The Meteoritical Society. (n.d.). *Search the meteoritical bulletin database*. www.lpi.usra.edu/meteor/

Thompson, W. L. (2016). *Living on the grid*. iUniverse.

Thunberg, G. (2019). *No one is too small to make a difference*. Penguin.

Tolkein, J. R. R. (1954–1955). *The fellowship of the ring; the two towers; the return of the king*. Allen & Unwin.

TOP500. (n.d.). *The list*. www.top500.org/

TradingEconomics. (n.d.). *Country list: Government debt to GDP*. https://tradingeconomics.com/country-list/government-debt-to-gdp

UIS. (n.d.). *Feature films and cinema data*. http://uis.unesco.org/en/topic/feature-films-and-cinema-data

UN. (2018). *World urbanization prospects: The 2018 revision*. United Nations, Dept Economic and Social Affairs, Population Division. www.un.org/en/events/citiesday/assets/pdf/the_worlds_cities_in_2018_data_booklet.pdf

UN. (2019). *World Population Prospects 2019*. https://population.un.org/wpp

UNESCO. (n.d.). *World heritage list*. https://whc.unesco.org/en/list

UNESCO. (2018). *World heritage sites: A complete guide to 1073 UNESCO world heritage sites* (8th ed.). Firefly.

UNGEGN. (2021). *Working group on exonyms*. http://ungegn.zrc-sazu.si/Home

UNODC. (2019). *Global Study on Homicide 2019*. UNODC. www.unodc.org/unodc/en/data-and-analysis/global-study-on-homicide.html

US Geological Survey. (2020). *Mineral Commodity Summaries 2020*. US Geological Survey.

Vaidhyanathan, S. (2017). *Intellectual property: A very short introduction*. OUP.

Vietmeyer, N. (2012). *Our daily bread: The essential norman borlaug*. Bracing.

VNS. (2018). *Vietnam faces double health burden: Malnutrition and obesity*. https://english.vietnamnet.vn/fms/society/210651/vietnam-faces-double-health-burden-malnutrition-and-obesity.html

Volcker, P. (2009, December 13). The only thing useful banks have invented in 20 years is the ATM. *New York Post*. http://nypost.com/2009/12/13/the-only-thing-useful-banks-have-invented-in-20-years-is-the-atm/

Wall, P. (1999). *Pain: The science of suffering*. Orion.

Watson, I. (1999). *The rule of the road, 1919–1986*. Rutgers University qualifying paper. www.ianwatson.org/rule_of_the_road.pdf

Webb, J. L. A., Jr. (2009). *Humanity's burden: A global history of malaria*. CUP.

Wellcome Trust. (2019). *Wellcome Global Monitor 2018*. Wellcome Trust. https://wellcome.ac.uk/reports/wellcome-global-monitor/2018

WHO. (2010). Pacific islanders pay heavy price for abandoning traditional diet. *Bulletin of the World Health Organization, 88*(7), 481–560.

WHO. (2018). *Global Status Report on Road Safety 2018*. World Health Organization. www.who.int/violence_injury_prevention/road_safety_status/2018/en

WHO. (2019). *World malaria report 2019*. World Health Organization. www.who.int/publications/i/item/world-malaria-report-2019

WHO. (2020). *Global health estimates: Life expectancy and leading causes of death and disability*. World Health Organization. https://www.who.int/data/gho/data/themes/mortality-and-global-health-estimates

WHO. (2021). *Global health observatory data repository*. https://apps.who.int/gho/data/node.main.MHSUICIDE

Wilmott, T. (2008). *The Roman amphitheatre in Britain*. History Press.

Winchester, S. (2003). *Krakatoa: the day the world exploded*. HarperCollins.

WIPO. (n.d.). *Intellectual property statistics.* www.wipo.int/ipstats/en/

Worldometer. (n.d.). *Covid-19 coronavirus pandemic.* www. worldometers.info/coronavirus

Worek, M. (2011). *Nobel: A century of prize winners.* Firefly.

World Bank. (n.d.-a). *Literacy rate, adult total (% of people ages 15 and above).* https://data.worldbank.org/indicator/SE.ADT.LITR.ZS

World Bank. (n.d.-b). *Automated teller machines (ATMs) (per 100,000 adults).* https://data.worldbank.org/indicator/fb.atm.totl.p5

World Bank. (n.d.-c). *GDP per capita, PPP (current international $).* https://data.worldbank.org/indicator/NY.GDP.PCAP.PP.CD

World Bank. (n.d.-d). *Global links.* http://datatopics.worldbank.org/world-development-indicators/themes/global-links.html

World Bank. (n.d.-e). *GDP (current US$).* https://data.worldbank.org/indicator/NY.GDP.MKTP.CD

World Economic Forum. (2021). *Global Gender Gap Report 2021: Insight Report.* World Economic Forum.

WWF. (n.d.). *Madagascar.* www.worldwildlife.org/places/madagascar

Index

A

Adenosine triphosphate (ATP), 35
Afghanistan, 21, 44, 125, 127, 131, 138, 149, 162
Africa (continent of), 2, 14, 21, 65, 66, 79, 81, 106, 115, 117, 125, 129, 131, 142, 147, 151, 154, 157, 164, 166, 168, 170, 172, 176, 181, 195, 197, 202
Albania, 73, 166
Alcohol, 205
Algeria, 193
Allen, P., 54
Alzheimer disease, 151
Amazon rainforest, 28, 30
Ambrose, C.E.L. Sir, 162
Amnesty International, 52, 187, 203
Amphitheatres, 96, 108, 204
Amundsen, R., 10
Amundsen–Scott South Pole Station, 10
Anderson, B., 90
Anderson, P., 19, 203
Annan, K., 63
Antarctica, 1, 7, 10, 13, 16, 18, 19, 59, 89, 181
Antigua and Barbuda, 138, 162, 191
Apple Inc., 189, 202, 206
Aral Sea, 16, 100, 197, 206
Argentina, 59, 73, 94, 159
Arles amphitheatre, 96
Armenia, 42, 164
Arnold, K., 127
Aruba, 183, 191, 208
Asimov, I., 21, 57, 203, 204
Athens, 161
A Toroidal LHC ApparatuS (ATLAS), 113, 114, 205
Attenborough, D. Sir, 66
Auden, W.H., 168
Australia, 2, 26, 44, 48, 54, 59, 65, 71, 74, 79, 95, 108, 115, 131, 135, 146, 161, 170, 181, 187, 202
Austria, 24, 50, 119, 166
Automated teller machines (ATMs), 179, 206
Aviation, 48, 68, 71, 204
Azerbaijan, 4

B

Bahamas, 81, 159, 191
Bahrain, 54, 55, 61, 65, 181
Baikonur Cosmodrome, 4, 100
Balk, D., 90
Bangladesh, 42, 54, 61, 74, 140, 162, 164, 174
Banking, 179, 185
Barbados, 81, 106, 138
Beijing, 68, 123

Belarus, 24, 42, 74, 154, 164
Belgium, 46, 119, 147, 176
Belize, 135, 164
Benz, K., 52
Berlin, 57, 162
Bermuda, 42, 106, 162, 166
Bezos, J., 102
Bhutan, 127, 152, 164
Biodiversity, 28, 31, 81, 83, 94, 197
Björk, A., 170
Black, J., 33
Body mass index (BMI), 138
Boki, I., 164
Bolivia (Plurinational State of), 42, 106, 162, 164
Bolt, U., 159
Borges, J.L., 168
Borlaug, N., 81, 203
Botswana, 54
Bradman, D.G. Sir ("The Don"), 108
Brazil, 4, 28, 30, 46, 54, 59, 65, 70, 73, 74, 79, 83, 110, 159, 176
Brits, J.H., 13
Brout, R., 114
Bruneau, G., 21, 23, 203
Brunei Darussalam, 65, 176
Buenos Aires, 87
Bunin, I., 168
Burj Khalifa, 98
Burkina Faso, 23, 54, 147, 162
Burundi, 151, 164, 189, 195
Byron, Lord, 27

C

Caesar, J., 125
Calcutta, 87
Calder Hall, 74
Calendars, 125, 127, 205
Calment, J., 154
Cambodia, 42, 129, 164, 197, 199
Cameroon, 164
Campbell, K., 44
Camus, A., 131
Canada, 9, 14, 44, 59, 65, 74, 79, 131, 135, 170, 172, 181, 187
Cancer, 16, 140, 149, 151, 152
Cape Canaveral, 100
Capital punishment, 6, 52, 203
Carbon dioxide, 28, 63, 66, 71, 79, 87, 90, 202, 204
Carlsen, M., 166
Carter, J., 179
Caspian Sea, 14, 197
Cave of Hands, 94

Printed in the United States
by Baker & Taylor Publisher Services